ELECTRONICS

FUNDAMENTALS *for the* WATER & WASTEWATER MAINTENANCE OPERATOR SERIES

ELECTRONICS

FRANK R. SPELLMAN, Ph.D.
JOANNE DRINAN

CRC Press
Taylor & Francis Group
Boca Raton London New York

CRC Press is an imprint of the
Taylor & Francis Group, an **informa** business

Electronics

First Published 2001 by Technomic Publishing Company, Inc.

CRC Press
Taylor & Francis Group
6000 Broken Sound Parkway NW, Suite 300
Boca Raton, FL 33487-2742

First issued in paperback 2019

© 2001 by Taylor & Francis Group, LLC
CRC Press is an imprint of Taylor & Francis Group, an Informa business

No claim to original U.S. Government works

ISBN-13: 978-1-56676-958-7 (hbk)
ISBN-13: 978-0-367-39767-8 (pbk)

Visit the Taylor & Francis Web site at
http://www.taylorandfrancis.com

and the CRC Press Web site at
http://www.crcpress.com

Library of Congress Catalog Card No. 00-108906

Main entry under title:
    Fundamentals for the Water and Wastewater Maintenance Operator: Electronics

Contents

1 INTRODUCTION

2 BASIC EQUATIONS AND OPERATIONS

3 DIRECT CURRENT (D-C) REVIEW

4   THE PN JUNCTION DIODE

5   TRANSISTORS

6   ALTERNATING CURRENT (A-C) REVIEW

7   RESONANCE

9 OSCILLATORS

10 POWER SUPPLIES

11 TRANSFORMERS

12 FINAL REVIEW EXAMINATION

APPENDIX A

APPENDIX B

INDEX

Series Preface

Currently several books address broad areas of wastewater and waterworks operation. Persons seeking information for professional development in water and wastewater can locate study guides and also find materials on technical processes such as activated sludge, screening and coagulation. What have not been available until now are accessible treatments of each of the numerous specialty areas that operators must master to perform plant maintenance activities and at the same time to upgrade their knowledge and skills for higher levels of certification.

The *Fundamentals for the Water and Wastewater Maintenance Operator Series* is designed to meet the needs of operators who require essential background knowledge of subjects often overlooked or covered superficially in other sources. Written specifically for maintenance operators, the series comprises focused books designed to enhance knowledge and understanding.

Fundamentals for the Water and Wastewater Maintenance Operator Series covers over a dozen subjects in volumes that form stand-alone information guides or elements of a library of key topics. Areas to be presented in series volumes include: electricity, electronics, water hydraulics, water pumps, handtools, blueprint reading, piping systems, lubrication, and data collection.

Each volume in the series is written in a straightforward style without jargon or complex calculations. All are heavily illustrated and include extensive, clearly outlined sample problems. Self-check tests are found within every chapter, and a comprehensive examination concludes each book.

The series provides operators with the information required for improved job performance. Equally important, using key points, worked problems, and sample test questions, the series is designed to help operators answer questions and solve problems on certification and licensure examinations.

Preface

The effect of changes in the field of electronics is most striking in the area of consumer goods. One has only to compare a shopping trip in 1959 to a similar excursion in 2000. Forty years ago, electronics was governed by the vacuum tube, with transistors just coming into vogue. Devices were bulky and balky—TVs the size of easy chairs with mechanically moved aerials and phonographs spinning vinyl records.

Fast forward to a big-box electronics warehouse, where personal computers, miniature televisions, portable stereos, pagers and games are displayed, and you are easily persuaded of the microchip revolution, at least in comparison to what was available not so long ago.

As the technology of electronics has advanced, so have the training requirements. Unfortunately, the information available and the ways of presenting it have not often kept pace with specialists who must work with electronic equipment, a group that includes water and wastewater operators.

Many water and wastewater operators are given minimal training in electronics, and what they do receive often leaves them believing the subject is too difficult or mysterious, when actually the opposite is true, as this book will demonstrate.

Electronics has pared down the study of electronics to the important principles that need to be known by operators working with electronic equipment. The book avoids jargon and presents information in clear terms that are readily applicable to plant situations. While the book is practical in tone, it is not intended to transform operators into electronics technicians. Instead, its purpose is to present basic electronic information so that operators can better understand the functioning of plant equipment and appreciate the role of electronics in plant analysis and control. The information found in these pages will also be helpful in answering questions on certification examinations.

To assure correlation to modern practice and design, illustrative problems are presented in terms of commonly used circuit parameters (voltages, currents etc.) and cover those circuits and devices typically found in today's electronic systems.

The text is accessible to those with no experience in electronics. However, an understanding of basic electrical principles will help the reader fully appreciate this volume. Operators requiring such understanding may wish to consult another volume in the *Fundamentals for the Water and Wastewater Maintenance Operator Series* titled *Electricity.*

Each chapter ends with a Self-Test to help evaluate your mastery of the concepts presented. Before going on to the next chapter, take the Self-Test, compare your answers to the key, and review information for any problems missed. A comprehensive final exam can be found at the end of the text.

IMPORTANT   **Note:** *The symbol* ✔ *(check mark) displayed in various sections throughout the manual indicates a point is especially important and should be studied carefully.*

Acknowledgements

As with the other manuals in this series, Electronics fills a need specific to water/wastewater workers. Throughout this worldwide industry, water and wastewater operators have worked with, around, and (in some cases) on electrical/electronic equipment of all kinds. Waterworks operators in particular must know basic electrical/electronic principles and concepts for their jobs, and many licensure examinations test them on this information. Unfortunately, training in basic electricity/electronics has not always been effective, available, or afforded to operators who needed it. This particular manual fills this void. We'll continue to address needed information in several more volumes in this series, which covers often-neglected maintenance topics.

Simply put, this manual's objective is to bring together under one cover the most salient basic electronic theory, principles, practices, and processes as they should be addressed in a sound basic electricity/electronics training program. This manual provides a ready source of practical answers to the most common electrical/electronic problems facing those water/wastewater operators who are practicing, teaching, and learning basic electronics.

The water/wastewater professionals with whom we are in constant contact in our careers as authors, lecturers, trainers, seminar presenters, and consultants have been very generous in sharing their expertise and experiences with us. The problems they have encountered and solutions that worked have been a rich source of useful information. To all of them, we continue to owe a large debt of gratitude.

In particular, we acknowledge the extensive training given to us by the U.S. Navy. Both of us had the distinct honor and pleasure of having served extended active duty careers in the Naval Service, where we had the good fortune of being trained at the Navy's Electrical/Electronic Class A (basic), B (advanced), and several C (very advanced) schools, as well as several other technical schools. We agree that the classroom training and subsequent on-the-job-training we received while on active duty was far superior to any that we could have received elsewhere. For this, we are eternally grateful.

This manual is a direct result of the extensive training we received while in the Navy. In particular, however, we need to acknowledge the help provided to us by the main reference resource we used in compiling this particular manual: The Bureau of Naval Personnel's *Basic Electronics* (1971), available through Dover Publications, Inc., New York, N.Y.

Introduction

1.1 SETTING THE STAGE

The name *electronics* comes from the word *electron*—a very small, invisible quantity of electricity present in all materials. In terms of its many uses, electronics can be defined to include all applications involving the control of electricity in the solid semiconductor materials used for transistors and integrated circuits.

The transistor and its offspring (the integrated circuit and microprocessor) **are** modern electronics. In electrical terms, we can say that these revolutionary devices have "short-circuited" much of traditional electronic theory, revolutionized its practice, revamped the whole technology, and created several new paths for the future of electronics to take.

 Note: *Because this volume covers only modern basic electronics, we do not discuss vacuum tubes and their operation.*
IMPORTANT

Electronics may seem a difficult and puzzling subject to many of us. But we know it is not mysterious or difficult. It is, in real terms, the function or output of practical applications of the basic principles of electricity and magnetism. This manual is a first step in the direction of fulfilling a need for ground-level basic electronic information and/or preparing you for more advanced electronics training.

In *Electronics*, we abandon the traditional approach normally used to teach electronics. The approach we use is governed by the tenets of simplicity and factual clarity. Taking advantage of this simple, clear approach allows us to present only those areas actually needed in modern electronics rather than those found in traditional textbooks. Simply put, if you want traditional electronics, we recommend traditional methodology readily found in all other presentations of electronics and in traditional texts on the subject.

The arrangement and approach of this manual has been tested in the real world of water/wastewater operations—it has been used to train maintenance operators who needed formal training on the subject matter.

Why does the water/wastewater maintenance operator need formal but concise training on electronics? We answer this question by asking a few questions of our own: Have you looked around your plant recently? Have you noticed that not only is the lab equipment you use electronic, but also the means by which you transmit the data (via the computer)? Have you noticed that in an effort to down-size or to prevent privatization, your plant is installing more and more electronically controlled equipment, including sampling devices? Have you taken a licensure examination recently and discovered (in many states) that the exam includes questions that deal with electrical or electronic subject areas?

If you can answer yes to any of these questions, this manual is recommended reading. In today's world, the water/wastewater maintenance operator who lacks knowledge of basic electricity/electronics is severely hampered, like a surgeon not trained to properly use the scalpel.

Each chapter is followed by a self-test, which is used to gauge your progress through each lesson. We begin with the mathematical basics essential to understanding basic electronics. Math is an important part of electrical/electronic theory. However, the math we present throughout this manual is "bare-bones"—only the operations that must be learned and nothing beyond the first-year algebra level. If you can master the basic equations presented in Chapter 2, the material presented in subsequent chapters will be much easier to master. In addition, we avoid semiconductor physics completely in this manual, because these are not needed for the level of knowledge for which we are striving.

This manual focuses on how to apply the few basic principles that are the basis of modern electronic practice. So, again, we know that electronics is not a mysterious or difficult subject, through our own experience.

Basic Equations and Operations

TOPICS

Axioms
Solving Basic Equations
Verifying the Answer
Powers of 10 and Scientific Notation
Basic Equations and Electrical Formulas
Sequence of Operations
Roots

Key Terms Used in This Chapter

EQUATION	A statement that two expressions or quantities are equal in value.
UNKNOWN	The number, value, or quantity to be found in an equation.
EXPONENT	Indicates the number of times a base is multiplied by itself.
FORMULAS	Equations in which letters or symbols stand for given values.
SQUARE ROOT	Of a number is one of the two equal factors of that number.
POWER	A product obtained by using a base a certain number of times as a factor.

2.1 INTRODUCTION

Problems in electronics are solved by the manipulation of mathematical equations, or formulae, in which some things are known and some unknown quantity must be determined. For example, if voltage and resistance are known, current can be found by setting up and then solving the appropriate equation.

An *equation* is a statement that two expressions or quantities are equal in value. The statement of equality $5x + 2 = 12$ is an equation; that is, it is algebraic shorthand for "the sum of 5 times a number plus 2 is equal to 12." It can be seen that the equation $5x + 2 = 12$ is much easier to work with than the equivalent sentence. Indeed, if we did not "shorthand" equations and had to work entirely with written statements, finding solutions to many problems would be difficult.

In solving simple equations, it is helpful to consider an equation as similar to a balance (see Figure 2.1-(a). The equal sign tells us that the two quantities are in balance (that is, they are equal). Suppose that the numbers in this equations are expressed in pounds; if 2 pounds are removed from each pan, the pans will still be in balance. Figure 2.1-(a) can then be revised as indicated in Figure 2.1-(b). The corresponding equation would then become $5x = 10$, a relationship that can be expressed as "five times an unknown number is equal to 10." That unknown number is 2.

The preceding solution may be summarized in three steps:

1. $5x + 2 = 12$

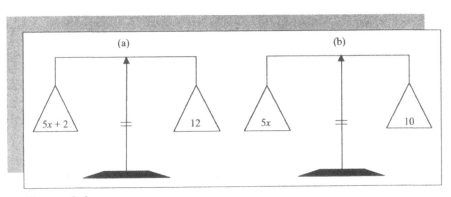

Figure 2.1
An equation is like a balance.

2. $5x = 10$

3. $x = 2$

Step 1 expresses the whole equation. In step 2, the number 2 has been subtracted from both sides of the equation. In step 3, both sides have been divided by 5.

An equation is, therefore, kept in balance (both sides are kept equal) by subtracting the same number from both sides, adding the same number to both, or dividing or multiplying both by the same number. Understanding this procedure, clearly, is the first step to successfully solving many equations and problems easily.

The number to be found in an equation is called the *unknown number*; the *unknown quantity*; or, simply, the *unknown*.

Solving an equation is finding the value or values of the unknown that make the equation true.

After completing this chapter, you will know how to (1) solve for an unknown quantity when any number of known quantities are given and (2) manipulate, or transpose, the terms of an equation to produce other useful equations. Knowing how to perform both of these operations makes working electronic problems much easier.

2.2 AXIOMS

An *axiom* is an established rule, principle, or law. Stated differently, an axiom is a truth accepted without proof. In solving the equations in Figure 2.1, two axioms were used (do you know which ones?). The following axioms are frequently used in the solution of equations; you should, therefore, study and understand each one:

1. If equal numbers are added to equal numbers, the sums are equal.

2. If equal numbers are subtracted from equal numbers, the remainders are equal.

3. If equal numbers are multiplied by equal numbers, the products are equal.

4. If equal numbers are divided by equal numbers (except zero), the quotients are equal.

5. Numbers that are equal to the same number or to equal numbers are equal to each other.

6. Like powers of equal numbers are equal.

7. Like roots of equal numbers are equal.

8. The whole of anything equals the sum of all its parts.

Note: *Axioms 2 and 4 were used to solve the equation $5x + 2 = 12$.*

IMPORTANT

2.3 SOLVING BASIC EQUATIONS

As we have already stated, solving an equation is determining the value or values of the unknown number or numbers in that equation.

EXAMPLE 2.1

Problem: Find the value of x if $x - 6 = 3$.

Solution:

Given equation:
$$x - 6 = 3$$
Add 6 to each side (axiom 1),
$$x = 3 + 6$$
Collect the terms (that is, add 3 and 6),
$$x = 9$$

IMPORTANT

Note: *Here we can see by inspection that x = 9, but inspection does not help in solving more complicated equations. But if you notice that to determine that x = 9, 6 is added to each member of the given equation, you have acquired an important method or procedure that can be applied to similar but more complex problems.*

EXAMPLE 2.2

Problem: Solve for x, if $x + 4 = 10$.

Solution:

Given equation:

$$x + 4 = 10$$

Subtract 4 from each side (axiom 2),

$$x = 10 - 4$$

Collect terms,

$$x = 6$$

EXAMPLE 2.3

Problem: Solve for c, if $5c = 25$.

Solution:

Given equation:

$$5c = 25$$

Divide each member by 5 (axiom 4),

$$c = 5$$

EXAMPLE 2.4

Problem: Solve for x, if $4x + 5 - 7 = 2x + 6$.

Solution:

Given equation:

$$4x + 5 - 7 = 2x + 6$$

Collect like terms $(+5)$ and (-7):

$$4x - 2 = 2x + 6$$

Add 2 to both members (axiom 1):

$$4x = 2x + 8$$

Subtract $2x$ from both members (axiom 2):

$$2x = 8$$

Divide both members by 2 (axiom 4):

$$x = 4$$

2.4 VERIFYING THE ANSWER

An equation always asks this question: What number must the unknown be for the two sides of the equation to be equal? The solution to the equation answers this question. After the solution is obtained, you should always verify it. You do this by substituting the solution for the unknown quantity in the given equation. If the two members of the equation are then identical, the number substituted is the correct answer.

EXAMPLE 2.5

Problem: Solve and verify $45x - 20 = 240 - 85x$

Solution:

$$45x - 20 = 240 - 85x$$
$$45x = 260 - 85x$$
$$130x = 260$$
$$x = 2$$

Substituting the answer $x = 2$ in the original equation,

$$45x - 20 = 240 - 85x$$
$$45(2) - 20 = 240 - 85(2)$$
$$90 - 20 = 240 - 170$$
$$70 = 70$$

Because the statement $70 = 70$ is true, the answer $x = 2$ must be correct.

2.5 POWERS OF TEN AND SCIENTIFIC NOTATION

IMPORTANT

Note: *In actual computational practice, the water/wastewater maintenance operator should realize that the accuracy of a final mathematical answer can never be better than the accuracy of the data used. Having stated the obvious, it should also be pointed out that correct and accurate data are worthless unless the maintenance operator is able to make correct computations.*

Two common methods of expressing a number—*powers of ten* and *scientific notation*—will be discussed in this section.

2.5.1 POWERS OF TEN NOTATION

An expression such as 5^7 is a shorthand method of writing multiplication. For example, 5^7 can be written as:

$$5 \times 5 \times 5 \times 5 \times 5 \times 5 \times 5$$

The expression 5^7 is referred to as *5 to the seventh power* and is composed of an *exponent* and a *base number*.

An exponent (or power of) indicates how many times a member is to be multiplied together. The base is the number being multiplied.

IMPORTANT

$$^7 \text{(exponent)}$$
$$5 \text{ (base)}$$

Let's look at base 5 again, using different exponents.

$$5^3$$

This is referred to as *5 to the third power*, or *5 cubed*. In expanded form,

$$5^3 = (5)(5)(5)$$

The expression *5 to the fourth power* is written as

$$5^4$$

In expanded form, this notation means

$$5^4 = (5)(5)(5)(5)$$

These same considerations apply to letters (a, b, x, y, etc.) as well. For example:

$$z^2 = (z)(z) \text{ or } z^4 = (z)(z)(z)(z)$$

When a number or letter does not have an exponent, it is considered to have an exponent of one. Thus:

IMPORTANT

$$5 = 5^1 \text{ or } z = z^1$$

The following examples help to illustrate the concept of powers notation.

EXAMPLE 2.6

Problem: How is the term 2^3 written in expanded form?

Solution: The power (exponent) of 3 means that the base number (2) is multiplied by itself three times:

$$2^3 = (2)(2)(2)$$

EXAMPLE 2.7

Problem: How is the term in.^{2} written in expanded form?

Solution: The power (exponent) of 2 means that the term (in.) is multiplied by itself two times:

$$\text{in.}^2 = (\text{in.})(\text{in.})$$

EXAMPLE 2.8

Problem: How is the term 8^5 written in expanded form?

Solution: The exponent of 5 indicates that 8 is multiplied by itself five times:

$$8^5 = (8)(8)(8)(8)(8)$$

EXAMPLE 2.9

Problem: How is the term z^6 written in expanded form?

Solution: The exponent of 6 indicates that z is multiplied by itself six times.

$$z^6 = (z)(z)(z)(z)(z)(z)$$

EXAMPLE 2.10

Problem: How is the term $(3/8)^2$ written in expanded form?

Solution:

When parentheses are used, the exponent refers to the entire term within the parentheses. Thus, in this IMPORTANT *example,*

$$(3/8)^2 = (3/8)(3/8)$$

When a negative exponent is used with a number or term, a number can be re-expressed using a posi- IMPORTANT *tive exponent:*

$$6^{-3} = 1/6^3$$

Another example is

$$11^{-5} = 1/11^5$$

EXAMPLE 2.11

Problem: How is the term 8^{-3} written in expanded form?

Solution:

$$8^{-3} = \frac{1}{8^3} = \frac{1}{(8)(8)(8)}$$

IMPORTANT

Note: *Any number or letter such as 3^0 or X^0 does not equal 3×1 or $X1$, but simply 1.*

EXAMPLE 2.12

When a term is given in expanded form, you can determine how it would be written in exponential form. For example,

$$(5)(5)(5) = 5^3$$

or

$$(in.)(in.) = in.^2$$

EXAMPLE 2.13

Problem: Write the following term in exponential form:

$$\frac{(3)(3)}{(7)(7)(7)}$$

Solution: The exponent for the numerator (remember: numerator/denominator) of the fraction is 2, and the exponent for the denominator is 3. Therefore, the term would be written as

$$\frac{(3)^2}{(7)^3}$$

IMPORTANT

Note: *It should be pointed out that because the exponents are not the same, parentheses cannot be placed around the fraction, and a single exponent cannot be used.*

EXAMPLE 2.14

IMPORTANT *It is common to see powers used with a number or term used to denote area or volume units (in.^{2}, ft^2, in.^{3}, ft^3) and in scientific notation.*

Problem: Write the following term in exponential form:

$$\frac{(\text{in.})(\text{in.})}{(\text{ft})(\text{ft})}$$

Solution: The exponent of both the numerator and denominator is 2:

$$\frac{(\text{in.})^2}{(\text{ft})^2}$$

The exponents are the same, which allows the term to be expressed as follows:

$$\left(\frac{\text{in.}}{\text{ft}}\right)^2$$

EXAMPLE 2.15

IMPORTANT *In moving a power from the numerator of a fraction to the denominator, or vice versa, the sign of the exponent is changed. For example,*

$$\frac{3^3 \times 4^{-2}}{8} = \frac{3^3}{8 \times 4^2}$$

2.5.2 SCIENTIFIC NOTATION

Scientific notation is a method by which any number can be expressed as a term multiplied by a power of 10. The term is always a term multiplied by a power of 10. The term is always greater than or equal to 1 but less than 10.

IMPORTANT

Examples of powers of 10 are

$$3.2 \times 10^1$$
$$1.8 \times 10^3$$
$$9.550 \times 10^4$$
$$5.31 \times 10^{-2}$$

The numbers can be taken out of scientific notation by performing the indicated multiplication. For example:

IMPORTANT

$$3.2 \times 10^1 = (3.2)(10)$$
$$= 32$$
$$1.8 \times 10^3 = (1.8)(10)(10)(10)$$
$$= 1800$$
$$9.550 \times 10^4 = (9.550)(10)(10)(10)(10)$$
$$= 95,500$$
$$5.31 \times 10^{-2} = (5.31)1/10^2$$
$$= 0.0531$$

An easier way to take a number out of scientific notation is to move the decimal point the number of places indicated by the exponent.

RULE 1

Multiply by the power of 10 indicated. A positive exponent indicates a decimal move to the *right,* and a negative exponent indicates a decimal move to the *left.*

EXAMPLE 2.16

Using the same examples above, the decimal point move rather than the multiplication method is performed as follows:

$$3.2 \times 10^1$$

The positive exponent of 1 indicates that the decimal point in 3.2 should be moved one place to the right:

$$3.2 = 32$$

The next example is:

$$1.8 \times 10^3$$

The positive exponent of 3 indicates that the decimal point in 1.8 should be moved three places to the right:

$$1.800 = 1800$$

The next example is

$$9.550 \times 10^4$$

The positive exponent of 4 indicates that the decimal point should be moved four places to the right:

$$9.5500 = 95,500$$

The final example is

$$5.31 \times 10^{-2}$$

The negative exponent of 2 indicates that the decimal point should be moved two places to the left:

$$05.31 = 0.0531$$

EXAMPLE 2.17

Problem: Take the following number out of scientific notation:

$$3.516 \times 10^4$$

Solution: The positive exponent of 4 indicates that the decimal point should be moved 4 places to the right.

$$3.5160 = 35{,}160$$

EXAMPLE 2.18

Problem: Take the following number out of scientific notation:

$$3.115 \times 10^{-4}$$

Solution:

$$0003.115 = 0.0003115$$

IMPORTANT

There are very few instances in which you will need to put a number or numbers into scientific notation, but you should know how to do it, if required. Thus, the method is discussed below. However, before demonstrating the process of putting a number into scientific notation, it is important to point out the procedure and the rule involved with the process.

Procedure: When placing a number *into* scientific notation, place a decimal point after the first non-zero digit. (Remember, if no decimal point is shown in the number to be converted, it is assumed to be at the end of the number). Count the number of places from the standard position to the original decimal point. This represents the exponent of the power of 10.

RULE 2

When a number is put *into* scientific notation, a decimal point move to the *left* indicates a *positive* exponent; a decimal point move to the *right* indicates a *negative* exponent.

Now, let's try converting a few numbers into scientific notation. First, convert 69 into scientific notation.

Note: *In order to obtain a number between 1 and 9, the decimal must be moved one place to the left. This move of one place gives the exponent, and the move to the left means that the exponent is positive:*

$$69 = 6.9 \times 10^1$$

Let's try converting another number.

$$1500$$

Remember, in order to obtain a number between 1 and 9, the decimal point must be moved three places to the left. The number of place moves (3) becomes the exponent of the power of 10, and the move to the left indicates a positive exponent:

$$1500 = 1.5 \times 10^3$$

Let's try a decimal number.

$$0.0661$$
$$0.0661 = 6.61 \times 10^{-2}$$

EXAMPLE 2.19

Problem: Put the following number into scientific notation:

$$6,969,000$$

Solution:

$$6,969,000 = 6.969 \times 10^6$$

EXAMPLE 2.20

Problem: Convert the following decimal to scientific notation:

$$0.000696$$

Solution:

$$0.000696 = 6.96 \times 10^{-4}$$

2.6 BASIC EQUATIONS AND ELECTRICAL FORMULAS

Basic equations in electricity/electronics start as **formulas**. For example, $E = I \times R$ (i.e., voltage = current $\times$ resistance; a basic Ohm's Law formula) is essentially an equation in which letters (or, in some cases, symbols) stand for given values. In the basic, but essential, Ohm's Law formula (i.e., $E = I \times R$), if any two values are known, the third can be found, but it may be necessary to develop a variation of the formula. If E and I are known, R may be found after the original formula is re-arranged.

$E = IR$ (divide through by I):

$\dfrac{E}{I} = \dfrac{IR}{I}$ (the Is on the right cancel out)

$\dfrac{E}{I} = R$

2.7 SEQUENCE OF OPERATIONS

An important rule in mathematics is always perform the indicated mathematical operation for a series of operations in the proper sequence, according to the applicable rules (note: Rules are designated by ✻ in the following). For example,

✱ **Rule:** In a series of additions, the terms may be placed in any order and grouped in any way.

Examples:

$$4 + 6 = 10 \text{ and } 6 + 4 = 10$$
$$(4 + 5) + (3 + 7) = 19, (3 + 5) + (4 + 7) = 19, \text{ and}$$
$$[7 + (5 + 4)] + 3 = 19$$

✱ **Rule:** However, in a series of subtractions, changing the order or the grouping of the terms may change the result.

Examples:

$$100 - 20 = 80, \text{ but } 20 - 100 = -80$$
$$(100 - 30) - 20 = 50, \text{ but } 100 - (30 - 20) = 90$$

✱ **Rule:** When no grouping is given, the subtractions are performed in the order written—from left to right.

Examples:

$$100 - 30 - 20 - 3 = 47$$
$$\text{or by steps, } 100 - 30 = 70, 70 - 20 = 50, 50 - 3 = 47$$

✱ **Rule:** In a series of multiplications, the factors may be placed in any order and in any grouping.

Examples:

$$[(3 \times 3) \times 5] \times 6 = 270 \text{ and } 5 \times [3 \times (6 \times 3)] = 270$$

✱ **Rule:** In a series of divisions, changing the order or the grouping may change the result.

Examples:

$$100 \div 10 = 10, \text{ but } 10 \div 100 = 0.1$$
$$(100 \div 10) \div 2 = 5, \text{ but } 100 \div (10 \div 2) = 20$$

✱ **Rule:** If no grouping is indicated, divisions are performed in the order written—from left to right.

Examples:

$$100 \div 5 \div 2 \text{ is understood to mean } (100 \div 5) \div 2$$

* Rule: In a series of mixed mathematical operations, the rule of thumb is whenever no grouping is given, multiplications and divisions are to be performed in the order written, then additions and subtractions in the order written.

EXAMPLE 2.21

Problem:

$$10 + 3 - 5 + 9 + 7 - 3 = ?$$

Solution: 21, by performing operations in the order in which they were given.

EXAMPLE 2.22

Problem:

$$120 \div 3 \times 5 \times 2 \div 2 = ?$$

Solution: 200, by performing the operations from left to right—in the order they occur.

EXAMPLE 2.23

Problem:

$$(12 \div 4) + (6 \times 2) - (6 \div 3)$$
$$+ (7 \times 3 \times 2) - 8 = ?$$

Solution:

$$3 + 12 - 2 + 42 - 8 = 47$$

First perform the multiplications and divisions and then the additions and subtractions.

Important Point: *Group the indicated operations as indicated in example 2.23.*

* **Rule:** In a series of different operations parentheses ( ) and brackets [ ] can be used to group operations in the desired order. Thus, $120 \div 3 \times 5 \times 2 \div 2 = [(120 \div 3) 5 \times 2] \div 2 = 200$.

2.8 ROOTS

Roots of numbers are the reverse of powers. In a product of equal factors, the repeated factor is called a **root** of the product. The *square root* of a number is one of the two equal factors of that number. Thus, 3 is the square root of 9 because $3 \times 3 = 9$. The *cube root* of a number is one of three equal factors of that number. The *fourth root* is one of the four equal factors, and so on for higher roots.

Square root operations are put to good use in many electrical and/or electronic calculations. For example, in the equation $Z^2 = R^2 + X^2$ to isolate Z on the left, we must take the square root of each side.

EXAMPLE 2.24

Problem: $R = 4$ and $X = 5$. Solve for Z using the formula $Z^2 = R^2 + X^2$.

Solution:

$$\sqrt{Z^2} = \sqrt{R^2 + X^2}$$
$$Z = \sqrt{4^2 + 5^2}$$
$$= \sqrt{16 + 25}$$
$$= \sqrt{41}$$
$$= 6.4$$

Self-Test

The answers to chapter self-tests are found in Appendix A.

2.1 $12 \div 3 + 8 \times 2 - 6 \div 2 + 7 = ?$

2.2 $16 \div 8 + 4 \times 2 \times 3 - 16 \times 2 \div 4 = ?$

2.3 $60 - 25 \div 5 + 15 - 100 \div 4 \times 2 = ?$

Solve the equations in 2.4 through 2.8.

2.4 $2x + 3 = x + 4$

2.5 $9x + 9 + 3x = 15$

2.6 $17x - 7x = x + 18$

2.7 $9y - 19 + y = 11$

2.8 $2y + 3y - 4 = 5y + 6y - 16$

2.9 Solve for R: $P = I^2R$

2.10 Solve for E: $I = E/R$

Direct Current (D-C) Review

TOPICS

Current Flow
Potential Difference (Voltage)
Resistance
Ohm's Law
Series and Parallel Resistive Circuits
Electric Power
E-I Graph
Kirchhoff's Laws
Voltage and Current Dividers
Switches
Capacitors

3.1 INTRODUCTION

Just as the foundation of a new house is usually constructed before the rest of the house is built upon it, the basics of electricity must be studied first before attempting to study electronics. This chapter is a review on those basic aspects of d-c (direct current) that apply to electronics. By no means does it cover the whole d-c theory (for more in depth coverage of d-c, we recommend the first volume of this series: *Electricity*), but merely those topics that are essential to basic electronics.

3.2 CURRENT FLOW

Electron movement, or flow, in a conductor is called *electric current*. To produce current, the electrons must be moved by a potential difference (or voltage). The flow of water is usually measured as the number of

Key Terms Used in This Chapter

AMPERE	The basic unit of electrical current.
CAPACITOR	Two electrodes or sets of electrodes in the form of plates, separated from each other by an insulating material called the dielectric.
CIRCUIT	The complete path of an electric current.
CONDUCTANCE	The ability of a material to conduct or carry an electric current. It is the reciprocal of the resistance of the material and is expressed in mhos.
CONDUCTOR	Any material suitable for carrying electric current.
DIRECT CURRENT (D-C)	An electric current that flows in one direction only.
ELECTRON	A negatively charged particle of matter.
FUSE	A protective device inserted in series with a circuit. It contains a metal that will melt or break when current is increased beyond a specific value for a definite period of time.
KIRCHHOFF'S LAWS	Forms the basis for d-c and a-c circuit analysis. **Current Law:** The summation of all currents at a junction equals zero. **Voltage Law:** The summation of all voltages in a loop equals zero.
LOAD	The power that is being delivered by any power-producing device. The equipment that uses the power from the power-producing device.
OHM'S LAW	A formulation of the relationship of voltage, current, and resistance, expressed as $E = IR$.
POTENTIAL	The amount of charge held by a body as compared to another point or body. Usually measured in volts.

POWER	The rate of doing work or the rate of expending energy. The unit of electrical power is the watt.
RESISTANCE	The opposition to the flow of current caused by the nature and physical dimensions of a conductor.
RESISTOR	A circuit element whose chief characteristic is resistance; used to oppose the flow of current.
RHEOSTAT	A variable resistor.
VOLT	The unit of electrical potential.
WATT	The unit of electrical power.

gallons per minute. Similarly, the number of electrons, or coulombs, that pass a point in a wire in 1 second measures the flow of electricity. While the flow of water does not have a special name, the flow of current does. This name is *ampere*, or amp, (A), which is the basic unit of measurement. Current is represented by the letter symbol I.

Note: The **coulomb** is the practical unit of electrical quantity that is equal to about 6 billion, billion electrons (6,000,000,000,000,000,000). The symbol for electrical quantity is Q.

IMPORTANT

In a conductor, such as copper wire, the free electrons are charges that can be forced to move with relative ease by a potential difference (i.e., a voltage). If a potential difference is connected across two ends of a copper wire (as shown in Figure 3.1), the applied voltage (in this case, 3 volts) forces the free electrons to move. This current flow is a movement of electrons from one end to the other, from the point of negative charge, $-Q$, at one end of the wire, moving through the wire, and returning to the positive charge, $+Q$, at the other end (analogous to how the motion in a whip is transmitted from one end to the other). The speed of this flow is very nearly equal to the speed of light, which is about

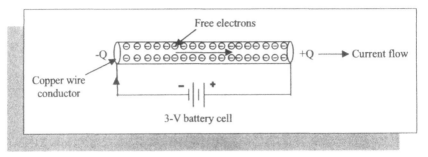

Figure 3.1
Electron (current) flow in a copper wire.

186,000 miles per second. The direction of electron flow is from the negative side of the 3-volt battery, through the wire, and back to the positive side of the battery. The direction of electron flow is from a point of negative potential to a point of positive potential.

Important Point: *We have defined electric current as a flow of electric charge. The electric charge consists of negatively charged electrons; however, with semiconductors, there will also be positive charge carriers called* **holes.**

3.2.1 ELECTRON FLOW VS. CONVENTIONAL CURRENT FLOW

Most of us are familiar with Benjamin Franklin and some of the experiments he conducted. Specifically, through experimentation with electricity, Franklin thought that electricity came from the positive terminal, or side, of the battery. Thus, he described an electric current flowing from the positive terminal, through the circuit, and back to the negative terminal. Figure 3.2 shows the current going from the positive battery connection, around the circuit, and back to the negative terminal. This is referred to as *conventional current flow*. Today, many involved with modern electrons follow this convention. This is true in many instances, because much of the mathematics of modern electronics follows this convention via agreement among engineers. They say current flows from positive to negative.

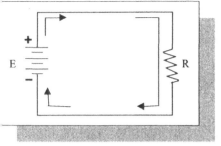

Figure 3.2
Conventional current; positive to negative.

Many of us were taught the opposite of conventional current flow theory. Instead, we were taught *electron flow theory*—that negatively charged electrons make up electrical current. In fact, today, we know (scientists know) that electrons flow from the negative terminal, through the circuit, and back to the battery. Figure 3.3 shows the direction of this **electron flow** through a typical, basic circuit. [**Note:** Although any circuit can be analyzed either with electron flow or by conventional current in the opposite direction, in this manual, we observe electron flow rather than conventional in our explanations and circuits. In those cases where this can't be done (for purposes of easier understanding), we clearly indicate which of these conventions we use].

Important Note: *Normally, we would ignore the old conventional current flow theory, and often we do. However, for those learning the basics of electronics, it is important to know about both conventions, because engineers still use conventional flow theory for many calculations. For example, in numerous texts, you will see that many semiconductor devices have a symbol that contains an arrowhead pointing in the direction of conventional current flow—positive to negative. Conventional current flow also provides a convenient mathematical solution to many problems. In this*

IMPORTANT

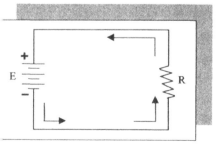

Figure 3.3
Electron flow; negative to positive.

manual, we are very careful to distinguish between conventional current and electron flow. Not all publications are as careful, however. As you study other books to learn more about electronics, be sure you understand whether the book means electron flow or conventional current when it refers to current flow.

In heavy industry, it is not unusual to find electrical equipment such as large horsepower motors that draw high amperages (this is especially the case at starting, when large motors can draw six to eight times normal running current). Some thermal equipment, such as ovens and dryers, draw several hundred amps. In most electronic equipment and/or applications, however, only small currents (fractions of an amp) are found. Amperage rated in milliamperes or microamperes is quite common. A *milliampere,* abbreviated mA, is one-thousandth of an ampere (i.e., 1/1000 or 0.001 amps). A *microampere*, abbreviated µA, is one-millionth of an amp (i.e., 1/1,000,000 or 0.000001 amps).

3.3 POTENTIAL DIFFERENCE (VOLTAGE)

In order to move water or air, a mechanical pressure supplied by a water pump (or gravity, i.e., difference in height) or an air compressor is required. Likewise, regardless of whether we term current flow conventional or electron flow current, in order to move electrons along a conductor or wire, electron pressure is required. This electrical pressure is called *potential difference* or *voltage* and is measured in units of *volts* (V).

IMPORTANT

Key Point: *The volt measures the amount of pressure, or push, in an electric circuit. This push is the force that moves electricity, so we sometimes refer to a voltage as an electromotive force, abbreviated Emf. Another name we often use for voltage is electrical potential, or just potential. The key concept to remember is that the purpose of all voltages is to produce a force to **move** electrons.*

Figure 3.4 sums up three important concepts related to potential difference (difference in potential or voltage) in an electrical circuit.

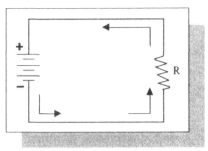

Figure 3.4
Difference in potential is supplied by the battery to the circuit.

1. The battery symbol indicates that a difference of potential is being supplied to the circuit.

2. The battery causes current to flow if there is a complete circuit present as shown in the figure.

3. The current will flow from negative to positive.

IMPORTANT

Note: *Even though the battery depicted in Figure 3.4 is a standard source of voltage and is used in most of the simple examples in this book as the source providing a potential difference, it is important to point out that there are other ways in which potential difference or voltage are generated. The main ways are chemically (battery), magnetically (generators or alternators), thermally, piezoelectrically, and photoelectrically. In practice, the most common ways used to produce potential difference are chemically and magnetically.*

3.4 RESISTANCE

From Figure 3.4, we learned that free electrons (current flow) may be forced to move when a voltage pressure (voltage) is applied. Different materials vary in their number of free electrons and in the ease in which they flow. A *conductor* is a material through which electrons may travel freely. Most metals (such as silver, copper, and aluminum) are good conductors. An *insulator* is a material that prevents the electrons from flowing through it easily. Nonmetallic materials like glass, mica, porcelain, rubber, and some textiles are good insulators. No material is a perfect insulator or a perfect conductor.

The ability of a material to resist the flow of electrons is called its *resistance*, is represented by the symbol *R*, and is measured in units called *ohms* (Ω). One ohm is defined as the amount of resistance that will limit the current in a conductor to one amp when the voltage applied to the conductor is one volt. Contrary to resistance, the ability of a material to conduct or carry an electric current is known as *conductance*. It is the reciprocal of the resistance of the material and is expressed in mhos.

Resistors are common components of many electrical and electronic devices. Some frequent uses for resistors are to establish the proper value of circuit voltage, to limit current, and to provide a load.

3.5 OHM'S LAW

Ohm's Law is the most basic equation in electricity. Ohm's Law states [see Equation (3.1)] the fundamental relationship between voltage, current, and resistance.

$$E = I \times R \qquad (3.1)$$

where

E = voltage, V
I = current, A
R = resistance, Ω

The basic Ohm's Law Equation (3.1) can be rearranged to find I and R as shown below:

$$I = E/R \qquad (3.2)$$

$$R = E/I \qquad (3.3)$$

The bottom line: If you know any two of the quantities E, I, and R, you can calculate the third.

Note: *Some textbooks state Ohm's Law as V = IR. V and E are both symbols for voltage. E is generally used for sources, and V is used for* IMPORTANT *voltage drops or differences in potential as well as for sources. In this manual, we use E for voltage sources and V for voltage quantity. Also, in this formula (i.e., E = IR), resistance is the opposition to current flow. Note that when resistance is large, the current will be small.*

EXAMPLE 3.1

Problem: Calculate voltage (*E*) for 0.002 amps through 1000 ohms of resistance, Ω.

Solution:

$$E = IR$$
$$E = 0.002 \text{ A} \times 1000 \text{ } \Omega$$
$$E = 2 \text{ volts}$$

EXAMPLE 3.2

Problem: Calculate current (*I*) for 12 volts applied across 8 Ω.

Solution:

$$I = E/R$$
$$I = 12 \text{ V}/8 \text{ } \Omega$$
$$I = 1.5 \text{ amps}$$

EXAMPLE 3.3

Problem: Calculate resistance (*R*) for 12 volts with 0.006 A.

Solution:

$$R = E/I$$
$$R = 12 \text{ V}/0.006 \text{ A}$$
$$R = 2,000 \text{ } \Omega \text{ or } 2 \text{ k}\Omega$$

Unlike amperage, resistance normally encountered in electronics is quite high—thousands to millions of ohms are used. The kilohm and megohm values are commonly used. The *kilohm* (k = kilo, Ω) measures resistance in thousands of ohms. Thus, 1 kΩ = 1000 ohms, 2 kΩ = 2,000 ohms, and 6.9 kΩ = 6900 ohms. The *megohm* (M = mega, Ω = ohm) measures resistance in millions of ohms. Thus, 1 MΩ = 1,000,000 ohms, and 6.3 MΩ = 6,300,000 ohms.

3.6 SERIES AND PARALLEL RESISTIVE CIRCUITS

A *series circuit* is a circuit in which there is only one path for current flow. As shown in Figure 3.5, the current I is the same in all parts of the circuit. This means that the current flowing through R_1 is the same as the current through R_2, is the same as the current through R_3, and is the same as the current supplied by the battery.

Total circuit resistance (R_T) in a series circuit is calculated using Equation (3.4).

$$R_T = R_1 + R_2 + R_3 \ldots R_n \qquad (3.4)$$

Note: *Total resistance is often referred to as the equivalent series resistance—R_{eq}.*

IMPORTANT

Resistors can also be connected in parallel as shown in Figure 3.6. From Figure 3.6, it is obvious that a *parallel circuit* is a circuit in which

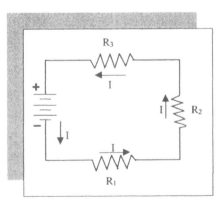

Figure 3.5
A series circuit.

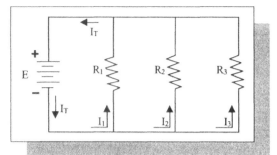

Figure 3.6
A parallel circuit.

two or more components are connected across the same voltage source. The resistors R_1, R_2, and R_3 are in parallel with each other and with the battery. Each parallel path is then a branch with its own individual current. When the total current I_T leaves the voltage source E, part I_1 will flow through R_1, part I_2 will flow through R_2, and the remainder I_3 through R_3. The point here is that the branch currents can be different. However, for voltage, the story is different. For example, if a voltmeter is connected across the individual resistors, the respective voltage drops will be equal. Therefore, in parallel circuits

$$E = E_1 = E_2 = E_3 \qquad (3.5)$$

The total current I_T is equal to the sum of all branch currents as shown in Equation (3.6). This equation applies for any number of parallel branches whether the resistances are equal or unequal.

$$I_T = I_1 + I_2 + I_3 \qquad (3.6)$$

The total resistance R_T in a parallel circuit is given by the formula

$$\frac{1}{R_T} = \frac{1}{R_1} + \frac{1}{R_2} + \frac{1}{R_3} + \cdots + \frac{1}{R_n} \qquad (3.7)$$

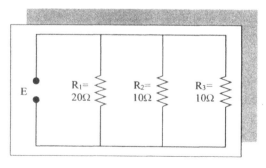

Figure 3.7
For Example 3.4.

EXAMPLE 3.4

Problem: Using the values in Figure 3.7, determine R_T.

Solution:

$$\frac{1}{R_T} = \frac{1}{20} + \frac{1}{10} + \frac{1}{10}$$

$$\frac{1}{R_T} = \frac{1}{20} + \frac{2}{20} + \frac{2}{20} = \frac{5}{20}$$

$$R_T = \frac{20}{5}$$

$$R_T = 4\,\Omega$$

Important Point: Note that R_T is smaller than the smallest of the resistors (10 Ω) in Example 3.4. This is an important point: R_T is **always** smaller than the smallest resistance in parallel.

IMPORTANT

3.7 ELECTRIC POWER

Electric power is generated by a source of applied voltage and is consumed in the resistance in the form of heat. Power is expressed in terms of **watts (w)**. The amount of power that the resistance dissipates in heat must be supplied by the voltage source; otherwise, it can't maintain the potential difference required to produce the current.

Note: Resistors used in electronics generally are manufactured in standard values with regard to resistance and power rating. Tables are available that provide standard resistance value. Quite often, when choosing a resistor for use in an electronic circuit, it is necessary to select a resistor with the closest standard value.

IMPORTANT

The electric power P used in any part of a circuit is equal to the current I in that part multiplied by the voltage V across that part of the circuit. Its formula is

$$P = EI \qquad (3.8)$$

where

P = power, w
E = voltage, V
I = Current, I

Other forms for $P = EI$ are

$$I = P/E \qquad (3.9)$$
$$E = P/I \qquad (3.10)$$

If we know the current I and the resistance R, but not the voltage V, we can find the power P by using Ohm's Law for voltage; substituting, we get

$$E = IR \qquad (3.11)$$

into Equation (3.8) we have

$$P = IR \times I = I^2R \qquad (3.12)$$

Likewise, if we know the voltage E and the resistance R but not the current I, we can find the power P by using Ohm's Law for current; substituting, we get

$$I = E/R \qquad (3.2)$$

into Equation (3.8) we have

$$P = E\frac{E}{R} = \frac{E^2}{R} \qquad (3.13)$$

As with any other like equation, if we know any two of the quantities, we can calculate the third.

EXAMPLE 3.5

Problem: The current through a 120-Ω resistor to be used in a circuit is 0.30 A. Find the power rating of the resistor.

Solution: Because I and R are known, use Equation (3.12) to find P.

$$P = I^2 R$$
$$P = (0.30)^2(120)$$
$$P = 0.09(120)$$
$$P = 10.8 \text{ w}$$

To prevent a resistor from burning out, the power rating of any resistor used in a circuit should be twice the wattage calculated by the power equation. Hence, the resistor used in this circuit should have a power rating of at least 22 w.

EXAMPLE 3.6

Problem: How many kilowatts of power are delivered to a circuit by a 220-V generator that supplies 22 A to the circuit?

Solution: *Because E and* I are given, use Equation (3.8) to find P.

$$P = EI$$
$$P = 220(22)$$
$$P = 4840 \text{ w} = 4.84 \text{ kw}$$

EXAMPLE 3.7

Problem: If the voltage across a 24,000-Ω resistor is 440 V, what is the power dissipated in the resistor?

Solution: Because R and E are known, use Equation (3.13) to find P.

$$
\begin{aligned}
P &= \frac{E^2}{R} \\
&= \frac{440^2}{24,000} \\
&= \frac{193,600}{24,000} \\
&= 8.07 \text{ w}
\end{aligned}
$$

3.8 E/I GRAPH

$I = E/R$ states that E and I are directly proportional for any one value of R. This relationship between E and I can be analyzed by using an E/I *graph* where the resistance is a fixed value.

In plotting an E/I graph (see Figure 3.8; the voltampere characteristic), the voltage values for E are marked on the horizontal axis (x axis or abscissa). The current values I are on the vertical axis (y axis or ordinate).

E and I depend on each other; thus, they are variable factors. The independent variable here is E because we assign values of voltage and note the resulting current. Usually, the independent variable is plotted on the x axis, which is why the E values are shown here horizontally while the I values are on the ordinate (y axis).

The two scales need not be the same. The only requirement is that equal distances on each scale represent equal changes in magnitude. In Figure 3.8, the x axis is stepped off in 2-V increments, while the

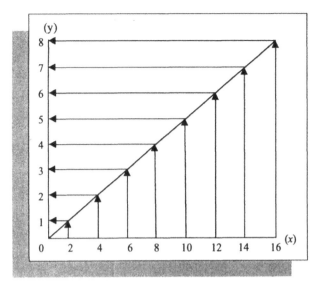

Figure 3.8
E/I graph.

y axis has 1-A incremental steps. The reference is the zero point at the origin.

The plotted points in the graph show the values to be placed (usually) in a table. A line joining these plotted points includes all values of *I*, for any value of *E*, with *R* constant.

It is important to point out that the straight line plotted in the graph shown in Figure 3.8 shows that *R* is a **linear** resistor (i.e., the resistor has a constant value of resistance; its *R* does not change with the applied voltage). Because *R* does not change with the applied voltage, then *E* and *I* are directly proportional. For example, doubling the value of *E* from 2 to 4 V results in twice the current, from 1 to 2 A. Similarly, four to five times the value of *E* will produce four to five times *I*, for a proportional increase in current.

The fact that the line plotted in Figure 3.8 is a straight line points to the linear nature of *R*. However, it is important to note that not all *E/I* plots result in straight lines. This is the case because of resistance that is **nonlinear**. As an example, the resistance of many temperature-sensitive resistive circuit components is nonlinear. That is, *R* increases with more current as the component becomes hotter. Increasing the applied voltage does produce more current, but *I* does not increase in the same proportion as the increase in *E*.

Note that whether R is linear or not, the current I is less for more R, with the applied voltage constant. This is an inverse relationship—I goes down as R goes up.

3.9 KIRCHHOFF'S LAWS

In a circuit where two voltages are applied in different branches, or in an unbalanced bridge circuit, the standard rules for series and parallel circuits can't be applied. In these cases (and others), more general methods of circuit analyses become necessary. These methods include the application of *Kirchhoff's Voltage and Current Laws*.

Kirchhoff's **voltage law** states that the *voltage supplied to a closed circuit must equal the sum of the voltage drops in the circuit*. This fact is expressed as follows:

$$E_A = E_1 + E_2 + E_3 \qquad (3.14)$$

where

E_A is the voltage and E_1, E_2, and E_3 are voltage drops.

Kirchhoff's voltage law can also be stated: *the algebraic sum of the voltage rises and voltage drops must be equal to zero.* A voltage source of emf is considered a voltage rise; a voltage across a resistor is a voltage drop.

 Note: *For convenience in labeling, letter subscripts are often shown for voltage sources and numerical subscripts for voltage drops. By transforming the right members of Equation (3.14) to the left side, Kirchhoff's voltage law can be written*

IMPORTANT

Voltage applied − sum of voltage drops = 0

Substitute letters:

$$E_A - E_1 - E_2 - E_3 = 0$$
$$E_A - (E_1 + E_2 + E_3) = 0$$

or

Using the Greek capital letter sigma [(Σ)—meaning sum of], we have

$$\sum E = E_A - E_1 - E_2 - E_3 = 0 \qquad (3.15)$$

Kirchhoff's Current Law states that the *sum of the currents entering a junction is equal to the sum of the currents leaving the junction.*

Sum of all currents entering = sum of all currents leaving

or

$$\sum I = 0 \qquad (3.16)$$

Kirchhoff's voltage and current laws are important electronics principles. They provide us with very powerful tools for studying electronics circuits and calculating conditions in those circuits.

3.10 VOLTAGE AND CURRENT DIVIDERS

Figure 3.9 shows a simplified circuit called a *voltage divider*, which is the basis for many major theoretical and practical concepts throughout the entire field of electronics. In simplest terms, the voltage drop R_2, E_O is the important point we want to stress here. The obvious question is how to determine the value of the voltage drop. This can be accomplished by using the following formula: (Keep in mind that $R_1 + R_2 = R_T$ is the actual resistance of the circuit)

$$E_O = E \times \frac{R_2}{R_1 + R_2} \qquad (3.17)$$

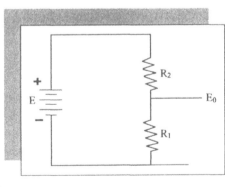

Figure 3.9
Voltage dividers.

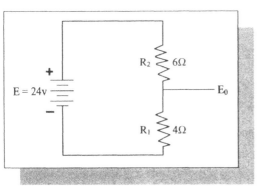

Figure 3.10
For Example 3.9.

EXAMPLE 3.8

Problem: Using Figure 3.10, what is the voltage drop (E_O) across R_2?

Solution:

$$E_O = E \times \frac{R_1}{R_2 \times R_1}$$

$$E_O = 24 \times \frac{4}{6 + 4}$$

$$E_O = 9.6 \text{ volts}$$

Note: *The output voltage from the voltage divider is always less*
*than the applied voltage. It is said to be **attenuated** (i.e., reduced;*
IMPORTANT *a term used extensively in communications equipment).*

Figure 3.11 shows a *current divider* branch circuit. The current in such a network splits or divides the two resistors that are in parallel with each other.

Figure 3.11 shows that I_T splits into the individual currents I_1 and I_2 and then these recombine to form I_T. The following relationships are valid for this branch circuit.

● $E = R_1 I_1$

● $E = R_2 I_2$

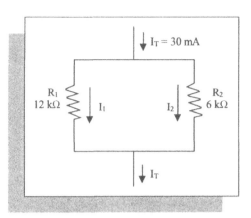

Figure 3.11
Current divider.

⬤ $R_1 I_1 = R_2 I_2$

⬤ $I_1 / I_2 = R_2 / R_1$

The following equations may be used to solve for branch currents:

$$I_1 = \frac{(I_T)(R_2)}{(R_1 + R_2)} \qquad (3.18)$$

$$I_2 = \frac{(I_T)(R_1)}{(R_1 + R_2)} \qquad (3.19)$$

EXAMPLE 3.9

Problem: Find I_1 and I_2 in Figure 3.11.

Solution:

1.

$$I_1 = \frac{(I_T)(R_2)}{(R_1 + R_2)}$$

$$I_1 = \frac{(0.03)(6)}{(12 + 6)}$$

$$I_1 = \frac{0.18}{18}$$

$$I_1 = 0.01 \text{ or } 10 \text{ mA}$$

2.

$$I_2 = \frac{(I_T)(R_1)}{(R_1 + R_2)}$$

$$I_2 = \frac{(0.03)(12)}{12 + 6}$$

$$I_2 = \frac{0.36}{18}$$

$$I_2 = 0.02 \text{ or } 20 \text{ mA}$$

3.11 SWITCHES

A *switch* is a device used in an electrical circuit for making, breaking, or changing connections under conditions for which the switch is rated. Switches are rated in amps and volts; the rating refers to the maximum voltage and current of the circuit in which the switch is to be used. In the **on** position, the closed switch has very little resistance. All the circuit current will pass through the switch because it is placed in series with the voltage source and its load. Open, the switch has infinite resistance, and no current flows in the circuit.

Many types and classifications of switches have been developed. The switch in Figure 3.12 is a single-pole single-throw (SPST) switch. It provides an **on** or **off** position for one circuit. Two connections are necessary.

As illustrated in the above example (shown in Figure 3.12), a common designation is by the number of poles, throws, and positions they have. The number of poles indicates the number of terminals at which current can enter the switch. The throw of a switch signifies the number of circuits each blade or contactor can complete through the switch. The number of positions indicates the number of places at which the operating device (toggle, plunger, knife, rocker or DIP, push-button, etc.) will

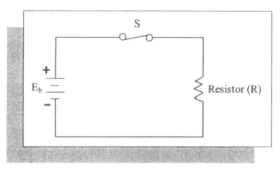

Figure 3.12
*Schematic diagram with general symbol for switch,
S; a single-pole, single-throw (SPST) type switch.*

come to rest. Figure 3.13 shows the schematic diagrams of some-often
used switches.

IMPORTANT

Important Points: *(1) When a switch is **closed** (or **on**), it has
total circuit current flowing through it; there is, however, **no**
voltage drop across its terminals. (2) When a switch is **open** (or
off), it has **no** current flowing through it; the full circuit voltage
appears across its terminals.*

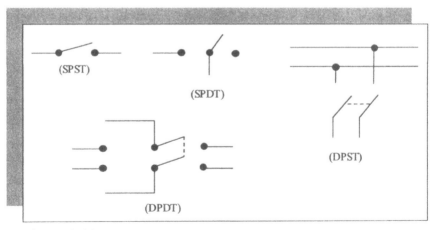

Figure 3.13
*Schematic diagrams of commonly used switches: SPST—single-pole, single-
throw; SPDT—single-pole, double-throw. DPST—double-pole, single-throw;
DPDT—double-pole, double-throw.*

3.12 CAPACITORS

A *capacitor* is an electrical component of two or more conductive plates separated by an insulating material. A capacitor stores energy in an electrostatic field.

Capacitors are used extensively in electronics, such as in tuning circuits, filter circuits, coupling circuits, bypasses for alternating currents of high frequency, and blocking devices in audio circuits. They are mainly used with a-c signals, but there are certain specific areas of d-c where they are very important. For example, in most electronic circuits, a capacitor has d-c voltage applied, combined with a much smaller a-c signal voltage. The usual function of the capacitor is to block the d-c voltage but pass the a-c signal voltage, by means of the charge and discharge current. In each application, the capacitor operates by storing up electrons and discharging them at the proper time.

 Note: *Practical applications of capacitor charging include preventing distortion in audio and communications circuits,* IMPORTANT *timing circuits, and in deliberately generating and altering electronic signals.*

Because their main function in d-c applications is to become charged and hold the charge, there are a few essential key points we need to know about capacitors used in d-c applications.

For instance, consider Figure 3.14 above. When the switch is closed, the capacitor will charge. When fully charged, the capacitor will charge to a final voltage charge of 12 volts. (Note: The capacitor will charge up

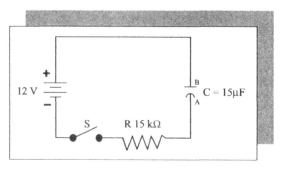

Figure 3.14
Capacitor used in a circuit with a d-c supply.

to the voltage that would appear across an open circuit located at the same place where the capacitor is located). For practical applications, the important question is, How long does it take to reach this voltage? To make this determination, we need to know the **time constant** of the circuit. The formula for time constant is:

$$T = RC \qquad (3.20)$$

where

T = Time Constant
R = Resistance
C = Capacitance

We can determine the time constant for the circuit shown in Figure 3.14 as follows:

$$T = RC$$
$$T = 12 \text{ k}\Omega \times 15 \text{ }\mu\text{F}$$
$$T = 0.18 \text{ seconds}$$

It takes approximately five time constants, or about 0.9 seconds, for the capacitor to reach 12 volts. In one time constant, the capacitor charges to 63% of the final voltage, or about 7.6 volts.

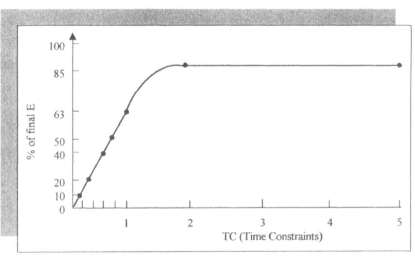

Figure 3.15
Capacitor charging graph.

In the circuit shown in Figure 3.14, the capacitor will remain un-charged until the switch is closed. [Note: The capacitor's plates, A and B (when charged or discharged), have the same voltage applied]. Before the switch is closed, the voltage on plate A and plate B both will be at 0 volts (if the capacitor is fully discharged). When the switch is closed, the volt-age on plate A will increase toward 12 volts, while the voltage on plate B will stay at 0 volts. After one time constant, the voltage on plate A will be about 7.6 volts.

Capacitor charges can be drawn on a graph (called an **exponential curve**) for the voltage rise plotted against time (see Figure 3.15).

3.12.1 CAPACITORS IN SERIES AND PARALLEL

Connecting capacitances in series is equivalent to increasing the thick-ness of the dielectric. Therefore, the combined capacitance is less than the smallest individual value. Figure 3.16 shows capacitances in series and equivalent.

The combined equivalent capacitance is calculated by the reciprocal formula:

$$\frac{1}{C_T} = \frac{1}{C_1} + \frac{1}{C_2} + \frac{1}{C_3} + \cdots + \frac{1}{C_n} \qquad (3.21)$$

The total capacitance of two capacitors in series is

$$C_T = \frac{C_1 C_2}{C_1 + C_2} \qquad (3.22)$$

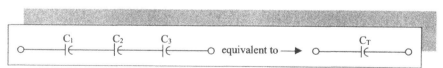

Figure 3.16
Capacitances in series.

EXAMPLE 3.10

Problem: What is the total capacitance and working voltage of a capacitance series combination if C_1 and C_2 are both 400 μF 200-volt capacitors?

Solution:

$$C_T = \frac{C}{n} = \frac{400}{2} = 200 \text{ μF}$$

The total working voltage that may be applied across a group of capacitors in series is equal to the sum of the working voltages of the individual capacitors. Therefore,

$$\text{Working Voltage} = 200 + 200 = 400 \text{ V}$$

Connecting capacitances in parallel (see Figure 3.17) is equivalent to adding the plate areas. Therefore, the total capacitance is the sum of the individual capacitances.

$$C_T = C_1 + C_2 + C_3 + \cdots + C_n \qquad (3.23)$$

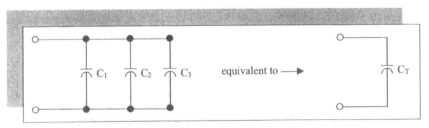

Figure 3.17
Capacitances in parallel.

EXAMPLE 3.11

Problem: What is the total capacitance of a capacitance series combination if $C_1 = 5$ μF, $C_2 = 10$ μF, and $C_3 = 20$ μF?

Solution:

$$C_T = \frac{1}{5} + \frac{1}{10} + \frac{1}{20} = 0.2 + 0.1 + 0.05$$
$$= 0.35$$
$$C_T = \frac{1}{0.35}$$
$$C_T = 2.86 \text{ μF}$$

Self-Test

3.1 Give the unit of measurement and symbol for each of the following: (a) difference in potential; (b) rate of current flow; (c) quantity of electricity; (d) resistance.

3.2 Changing what other two variables can change the rate of current flow in a circuit?

3.3 What three factors affect the resistance in a circuit?

3.4 Give two ways of increasing the resistance of the circuit wiring:

3.5 How much current is used to illuminate a 200-watt light bulb rated at 110 V?

3.6 What is the difference between a series and parallel circuit?

3.7 The sum of all voltage drops in a series circuit is equal to _____.

3.8 What is the total capacitance of a parallel combination of three 120-microfarad capacitors?

3.9 Define *RC* time constant.

3.10 How much is the *RC* time constant for 2 MΩ in series with 2 μF for charge?

The PN Junction Diode

TOPICS

Semiconductors
Diode Junction
Basic Junction Diode Operation
Diode Characteristic Curve
Diode Specifications
Zener Diodes
Varactor Diodes
Light-Emitting Diodes (LED)

4.1 INTRODUCTION

Recall that with a purely resistive circuit, it does not matter whether we pass d-c or a-c through it. The electron flow (current) through the resistive circuit depends on the applied voltage and the amount of resistance.

With an ideal capacitive circuit, however, we'll get different results. If we apply a-c voltage to the capacitor, an alternating current will flow. When we apply a d-c voltage, there will be a sudden impulse of current while the capacitor plates charge. After the capacitor is fully charged, however, there will be no current flow.

This chapter discusses devices that allow current to flow **in one direction only**. It doesn't matter which type of voltage we apply to these devices—current only flows in one direction. We refer to such devices as semiconductor *diodes*—used extensively in modern electronic circuits. For example, semiconductor diodes are commonly used in logic and decision circuits in computers and control circuits; as simple switching circuits that have no moving parts; for converting a-c to d-c; and for recovering the television and radio signals from those transmitted over the air, thus enabling us to see and hear the program.

Key Terms Used in This Chapter

SEMICONDUCTOR MATERIAL	A material with resistivity between that of metals and insulators. Pure semiconductor materials are usually doped with impurities to control the electrical properties.
DOPING	Adding impurities to pure semiconductor material to provide positive and negative charges.
JUNCTION DIODE	An electronic component formed by placing a layer of N-type semiconductor material next to a layer of P-type material. Diodes allow current to flow in one direction only.
FORWARD BIAS	A voltage applied across a semiconductor junction so that it will tend to produce current.
REVERSE BIAS	A voltage placed across a semiconductor junction so that it will tend to prevent current flow.
RECTIFIER CIRCUIT	Converts its a-c input to pulsating d-c output.
ZENER DIODE	Semiconductor diode used for voltage regulation. Takes reverse bias.

4.2 SEMICONDUCTORS

Most of the diodes in modern electronic circuits use semiconductor materials. This is the case because there are many advantages to using semiconductor diodes. For example,

● Semiconductors normally have low operating-voltage requirements.

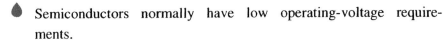

- They are rugged and reliable.
- They waste very little energy in the form of excess heat.

The obvious question is, What is a semiconductor?

Simply put, a *semiconductor* is a crystalline material with an electrical conductivity between that of metals (good) and insulators (poor). The conductivity of semiconductors can usually be improved by minute additions (**doping**) of substances or by other factors. Silicon (the most commonly used semiconductor material), for example, has poor conductivity at low temperatures, but this is improved by the application of light, heat, or voltage.

More specifically (avoiding the detailed physics of these materials), along with silicon, germanium, another semiconductor material, is also used in the manufacture of diodes and transistors (see Chapter 5). Both silicon and germanium are refined to an extreme level of purity, then minute, controlled amounts of a specific doping impurity is added.

Depending on the type of doping impurity used, the silicon or germanium semiconductor material is said to be N or P material. Adding impurities (e.g., antimony) that add electrons produces N-type material (named for the electron's negative charge). Impurities that add electrons are called **donor** materials—they donate an electron to the semiconductor material. Impurities (e.g., boron) that add holes are called **acceptor** materials; they produce P-type material (named for the hole's positive charge).

4.3 DIODE JUNCTION

When a piece of N silicon and a piece of P silicon are joined together, a PN junction (shown in Figure 4.1), normally called a *diode junction*, is formed. Diode junctions can also be made with N and P germanium. It is

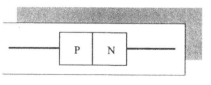

Figure 4.1
PN diode junction.

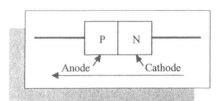

Figure 4.2
Current flow in diode.

important to point out, however, that silicon and germanium are never mixed together when making PN junctions.

The diode's main characteristic is that current will flow through it **in one direction only**. Figure 4.2 shows the direction in which the current flows.

IMPORTANT

Note: *In Figure 4.2, current flow is shown from cathode to anode (flow from negative to positive). This is based on the electron flow theory used in this manual. In texts where the conventional current flow theory is used, current flow is from anode to cathode (from positive to negative).*

The schematic symbol for a diode is shown in Figure 4.3. Again, the direction of current flow is against the arrowhead (electron flow theory), cathode to anode. The anode and cathode are indicated in the figure, but are not usually shown in circuit diagrams. Therefore, you must know which is which.

4.4 BASIC JUNCTION DIODE OPERATION

Figure 4.4 shows the structure of a PN junction diode. It is important to note that the two regions of the diode are not connected together mechanically. Instead, it is the behavior of the electrons and holes near the junction that allows current in only one direction.

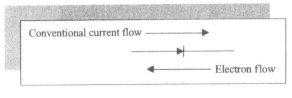

Figure 4.3
Schematic symbol for diode.

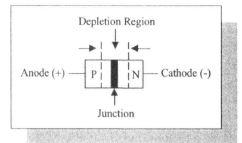

Figure 4.4
Structure of junction diode.

Figure 4.5 shows a PN junction diode connected to a d-c source. When connected to a d-c source as shown, the diode allows current through it. The battery voltage must be high enough to push (and pull) the charge carriers across the depletion region. (This takes about 0.3 volts for germanium and about 0.7 volts for silicon semiconductors). Once the electrons and holes cross the depletion region, the voltage will attract them the rest of the way through the semiconductor. The positive voltage attracts electrons from the N-type material all the way across the junction, through the P-type material. Then, the electrons travel through the circuit to the positive battery terminal.

In the same way, the negative battery terminal attracts the holes across the junction and through the N-type material; that is, holes flow in the opposite direction from electrons in a circuit. Notice the schematic symbol for a diode drawn next to the semiconductor block. With the negative voltage connected to the cathode and the positive voltage connected to the anode (i.e., with negative to negative and positive to positive), we

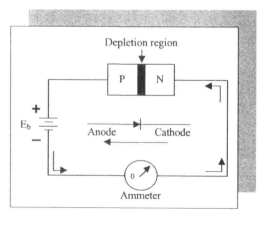

Figure 4.5
Forward biased diode circuit.

say the diode is *forward biased*. Additional bias can be applied from an external voltage source to provide control capability. The meter in the circuit shown in Figure 4.5 indicates some current through the circuit. In this example, the diode acts like a small resistance.

Note: *For every free electron that leaves the semiconductor on the P side, a new one comes onto the N side from the negative* IMPORTANT *battery terminal. Thus, one of the main purposes of the battery is to supply the electrons needed to maintain a current through the circuit. Because the semiconductor has been doped with either acceptor or donor electrons, when voltage is removed, the semiconductor material will keep as many electrons as it started with.*

If the battery terminals are reversed, as shown in Figure 4.6, the electrons move toward the positive terminal (unlike charges attract) and away from the junction (like charges repel), and the holes move toward the negative terminal (unlike charges attract) and away from the junction (like charges repel). The result is that the area near the junction, known as the *depletion region* (shown in Figure 4.6), gets wider because the holes and electrons are pulled away from the junction. There is no current because electrons and holes do not make it all the way through the semiconductor. The depletion region becomes too wide for them to cross. A diode that is connected in this way—with negative to positive and positive to negative—is said to be *reverse-biased*. In a reverse-biased diode, because

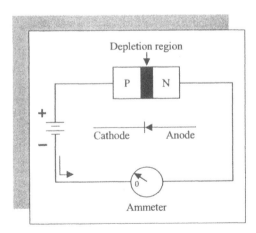

Figure 4.6
Reverse-biased diode circuit.

holes and electrons flow away from the junction, very few meet, and only a very small current flows (this very small current flow is indicated in Figure 4.6 by the ammeter indicator, which is not quite at the 0 level).

Why does a reverse-biased PN junction diode permit a small amount of current flow? The reverse-biased PN junction diode allows a small amount of current flow because of heat (thermal) energy, which continuously dislodges a very small number of electrons and holes near the junction, creating pairs of free electrons and holes in the depletion layer. However, the important point is that current is **approximately** zero in a reverse-biased diode. Again, external bias can be applied.

From the discussion to this point, you may have determined an important, practical application of the diode in an electronic circuit; that is, its functioning as a mechanical switch. This makes sense when you consider that the diode permits a current in one direction and prevents it in the other. Diodes, however, have two definite advantages over the common mechanical switch: (1) diodes can operate faster, because they are controlled by electric polarity; and (2) diodes generally last much longer, because they have no moving parts.

4.5 DIODE CHARACTERISTIC CURVE

The graph shown in Figure 4.7 illustrates a typical curve of a junction diode. The graph shows the two different kinds of bias: forward- and reverse-bias. Bias in the PN junction is the difference in potential between the anode (P material) and the cathode (N material). Recall that forward bias is the application of a voltage between N and P material, where the P material is positive with respect to the N material. When the P material becomes negative with respect to the N material, the junction is reverse-biased. Application of greater and greater amounts of forward bias causes more and more forward current until the power handling capability of the diode is exceeded, unless limited by external circuitry. Small amounts of forward bias cause very little current flow until the internal barrier potential is overcome. The potential difference varies from diode to diode, but is usually no more than a few tenths of a volt. Reverse bias produces a very small amount of reverse current until the breakdown point is

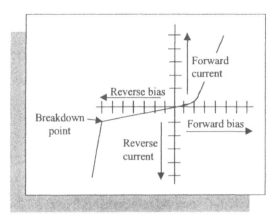

Figure 4.7
Diode characteristic curve.

reached, then an increase in reverse bias will cause a large increase in reverse current. Therefore, if breakdown is not exceeded, the ratio of forward current to reverse current is large—for example, milliamperes to microamperes or amperes to milliamperes. Changes in temperature may cause alterations in the characteristic curve, such as slope of curve at any point, breakdown point, amount of reverse current, etc. Note that when the diode is forward-biased and a small voltage is applied, the current is quite large.

> **Note:** *Characteristic curves like the one shown in Figure 4.7 can be seen on the screen of a curve trace (a special type of oscilloscope) when the scope is connected to a diode. The current through the diode, measured in milliamps, is plotted along the vertical line. The potential difference across the diode, measured in volts, is plotted along the horizontal line.*

IMPORTANT

4.6 DIODE SPECIFICATIONS

Manufacturers provide diode specification sheets and semiconductor data manuals to help in the selection of correct diodes for particular applications. Different ratings are given for different kinds of diodes. Descriptions of various ratings that are important to all diodes include the following

- **Average Forward Current (I_F)**—the maximum current that a diode can carry without being destroyed by heat. The diode, when forward-biased, generates power (resulting in heat) equal to the current times the voltage. The diode must be able to **get rid of** (dissipate) the heat safely. The temperature is normally specified for a range, typically $-65°$ to $+175°C$.

- **Average Forward Voltage (V_F)**—the voltage required at the desired forward current. It is greater than 0.7 V for silicon diodes.

- **Average Reverse Current (I_R)**—average reverse current that passes through a reverse-biased diode at a specified temperature. It is less than 1 μA for silicon diodes.

- **Reverse Breakdown Voltage or [Reverse Avalanche Voltage (V_{BR})]**—the reverse-bias voltage that causes the diode current to increase suddenly in the negative direction.

- **Reverse Recovery Time (t_{rr})**—time required for reverse current to decrease from a value equal to the forward current to a value equal to I_R when a step function of voltage is applied. Simply stated, it is the time needed to turn off the diode.

Note: *It is important to point out that the preceding diode specifications are normally rated at 25°C, because changes in temperature change the values of these specifications.*

IMPORTANT

4.7 ZENER DIODES

Zener diodes (named after C. A. Zener, who analyzed the voltage breakdown of insulators) are unique compared to other diodes in that they are designed to operate reverse-biased in the breakdown (or avalanche) region. They are also called **voltage-reference diodes** (or **breakdown diodes**, because they are designed for use in the breakdown region). The device is used mainly as a voltage regulator (when the avalanche current flows, the voltage across the diode remains constant), but is also used as

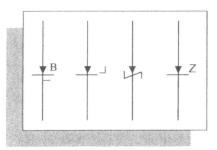

Figure 4.8
Zener diode schematic symbols.

a clipper, coupling device, and in other functions. Multiple diodes in series increase the voltage rating.

Schematic diagrams of the zener are shown in Figure 4.8. Zener current flows in the direction of the arrow. In many schematics, a distinction is not made for this diode, and a special diode symbol is used.

4.8 VARACTOR DIODES

PN junctions exhibit capacitance properties because the depletion area represents a dielectric and the adjacent semiconductor material represents two conductive plates. Increasing reverse bias decreases this capacitance while increasing forward bias increases it. When forward bias is large enough to overcome the barrier potential, high forward conduction destroys the capacitance effect, except at very high frequencies. Therefore, the effective capacitance is a function of external applied voltage. This characteristic is undesirable in conventional diode operation, but is enhanced by special doping in the *varactor* or variable-capacitance (varicap) diodes. Application categories of the varactor can be divided into two main types, tuning (e.g., television receivers) and harmonic generation. Solid-state varactors are simpler, smaller, and more stable than conventional capacitors, and they also last longer. Various schematic symbols are used to designate varactor diodes, as shown in Figure 4.9.

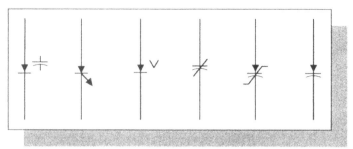

Figure 4.9
Schematic symbols of varactors.

4.9 LIGHT-EMITTING DIODES (LED)

A PN junction (i.e., the light-emitting diode—LED) radiates light when current passes through the unit; it looks like a tiny light bulb. The difference between the LED and a regular light bulb is the way the light is produced. In a standard light bulb, the filament (usually tungsten or other high-resistance filament wire) is heated by a current until the filament glows brightly. An LED produces radiation by allowing a forward current (to produce light it must be forward-biased) to pass through it in such a way that the electrons and holes in the semiconductor combine to produce visible light or radiation that is invisible to the human eye.

LEDs produce different colors of light—red, green, blue, yellow, or orange—depending on the semiconductor materials. The symbol used for an LED is shown in Figure 4.10. As with other diodes, the arrow in the LED symbol points in the opposite direction of electron current and in the direction of conventional current flow.

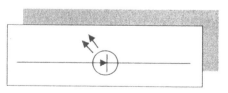

Figure 4.10
Schematic symbol for LED.

Self-Test

4.1 Define reverse bias.

4.2 Define forward bias.

4.3 Conventional current flow is from _____ to _____, while actual electron flow is from _____ to _____.

4.4 If diode is forward-biased, free electrons in the N-region and _____ in the P-region all move toward the junction.

4.5 If a diode's maximum forward current is exceeded, the diode will _____ and stop functioning.

4.6 Silicon is most conductive at _____ temperatures.

4.7 A junction diode consists of _____ semiconductor material.

4.8 What will occur if the positive terminal of a voltage source is connected to the N section of a junction of a diode?

4.9 What area in a junction diode has very few electron-hole pairs?

4.10 Why is silicon-type semiconductor material doped?

Transistors

TOPICS

Purpose of Transistors
Bipolar Junction Transistors (BJTs)
Transistor Connections
Transistor Characteristics
Field-Effect Transistors (FET)
The Junction Field Effect Transistor (JFET)
The Transistor as a Switch
Transistor Specifications
Integrated Circuits

5.1 INTRODUCTION

Just as the discovery of germ theory revolutionized medicine, the discovery of the transistor (and the transistor effect) revolutionized electronics.

Transistors have had a greater impact on the electronics industry than any other innovation to date—with thousands of millions now manufactured each year. John Bardeen and Walter Brattain, who were developing the work of the U.S. physicist William Shockley, first invented transistors at Bell Telephone Laboratories in the U.S. in 1948. Their useful qualities were immediately apparent. Transistors perform many of the functions of thermionic valves (vacuum tubes), but have the advantage of greater reliability, long life, compactness, and instantaneous action (no warming-up necessary). Widely used in most electronics equipment including portable radios and television, computers, and satellites, they are the basis of the integrated circuit (silicon chip).

In this chapter, we introduce the most common transistor type, commonly called the BJT, or bipolar junction transistor, as well as the JFET, or the junction field effect transistor. This chapter and later chapters concentrate on the basic operation of the BJT, because it is the most common.

Along with our discussion of transistor types and operating principles, we examine some of the many functions transistors can perform. In

Key Terms Used in This Chapter

DEPLETION MODE
Type of operation in a JFET where current is reduced by reverse bias on the gate.

DEPLETION REGION
An area around the semiconductor junction where the charge density is very small. This creates a potential barrier for current flow across the junction. In general, the region is thin when the junction is forward biased and becomes thicker under reverse-bias conditions.

DRAIN
The point at which the charge carriers exit an FET.

FIELD-EFFECT TRANSISTOR (FET)
A voltage-controlled semiconductor device. Varying the input voltage can vary output current. The input impedance of an FET is very high.

GATE CONTROL
Terminal of an FET.

INTEGRATED CIRCUIT
A device composed of many bipolar or field-effect transistors manufactured on the same chip, or wafer, of silicon.

JUNCTION FIELD-EFFECT TRANSISTOR (JFET)
A field-effect transistor created by diffusing a gate of one type of semiconductor material into a channel of the opposite type of semiconductor material.

N-TYPE MATERIAL
Semiconductor material that has been treated with impurities to give it an excess of electrons. We call this donor material.

PN-JUNCTION
The contact area between two layers of opposite-type semiconductor material.

SOURCE
The point at which the charge carriers enter an FET.

this chapter, we introduce the transistor as a switch. Later, in Chapter 7, we learn how a transistor can be made to operate as an amplifier.

5.2 PURPOSE OF TRANSISTORS

A *transistor* is a device that utilizes a small change in current to produce a large change in voltage, current, or power. The transistor, therefore, may function as an amplifier or an electronic switch. An amplifier is a device that increases the voltage, current, or power level of a signal applied to that device (see Chapter 7). When used as a switch, the transistor utilizes a small voltage or current to turn on or turn off a large current flow, with great rapidity.

Used as discrete system components, transistors are being used increasingly in integrated circuits (ICs). ICs, many of which contain thousands of transistors, make today's modern computers possible.

5.3 BIPOLAR JUNCTION TRANSISTORS (BJTs)

In Chapter 4, we pointed out that a junction diode is a two-terminal device composed of a P-type material and an N-type material. A transistor is also composed of layers of P-type and N-type materials, but is a three-terminal device (see Figure 5.1). (Note: Where there are

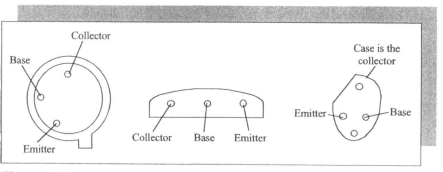

Figure 5.1
BJT three-terminal configurations.

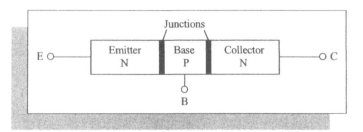

Figure 5.2
NPN transistor.

only two leads, the case is used instead, as shown in Figure 5.1 on the bottom).

Most transistors are composed of the semiconductor material silicon, although germanium is also sometimes used. The two basic types of transistors are

● *bipolar junction transistors* (BJTs), current-controlled devices (discussed in this section)

● *field-effect transistors* (or FETs), voltage controlled devices (discussed in Section 5.6)

Bipolar junction transistors (BJTs) are normally divided into two groups according to the arrangement of the semiconductor material:

● An *NPN* transistor, shown in Figure 5.2, consists of two layers of N-type material separated by a thin layer of P-type material.

● A *PNP* transistor, shown in Figure 5.3, consists of two layers of P-type material separated by a thin layer of N-type material.

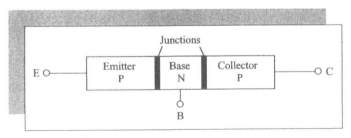

Figure 5.3
PNP transistor.

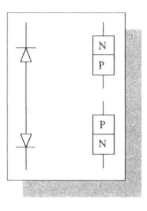

Figure 5.4
Simplest form of transistor: two diodes connected back-to-back.

In both the **NPN** and **PNP** transistors, one end region is called the *emitter*, the center region is called the *base*, and the other end region is called the *collector*. In its simplest form, a transistor can be considered as two diodes, connected back to back, as shown in Figure 5.4. [Note: The transistor has two junctions (see Figures 5.2 and 5.3), one located between the emitter and the base, and the other between the collector and the base].

At this point, the obvious assumption might be that a transistor is nothing more than two junction diodes connected back to back, as shown in Figure 5.4. Not at all.

Two separate diodes wired back to back *will not* behave like a transistor. This is because of semiconductor physics, which is beyond the scope of this manual. However, it is within the scope of this manual to point out that in the actual construction process, one very important modification is made to the arrangement shown in Figure 5.4. Instead of two separate P regions as shown, only one very thin region is used, as shown in Figure 5.5.

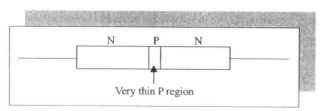

Figure 5.5
Instead of two P regions as shown in Figure 5.4, in practice, only one very thin P region is used.

Let's leave the junction diode and its back-to-back combination and move on to the NPN transistor shown in Figure 5.2. In the NPN transistor, the emitter is made of **N**-type material heavily doped with donor materials (the free electron supplier, for example, arsenic, antimony, or phosphorus). The free electrons in the emitter flow into the base, which is made of P-type material lightly doped with acceptor materials (which have positive holes that accept free electrons, for example, gallium, boron, or indium). The base is very narrow and lightly doped so that most of the electrons pass through this region into the collector, which is made of N-type material.

In the PNP transistor, shown in Figure 5.3, the emitter is made of **P**-type material with an excess of holes. Most of the holes pass through the base into the collector. The emitter is heavily doped with gallium, boron, or indium. The base is lightly doped with arsenic, antimony, or phosphorus.

5.3.1 BJT: BIASING AND BASIC CURRENT PATHS

Consider the BJTs shown in Figure 5.6 where the emitter-base junction is forward-biased. This allows the majority carriers—electrons in **NPN** transistors, holes in **PNP** transistors—in the emitter region to flow quite easily across the emitter-base junction. Once they are in the base, the majority carriers become the minority carriers. Note that the base-collector junction is reverse-biased—so that the minority carriers can flow into the collector. The movement of positive and negative charges for NPN and PNP transistors is shown in Figure 5.6.

Important Point: *On schematic diagrams, the letter Q always represents transistors. The schematic symbols for NPN and PNP* IMPORTANT *transistors are shown in Figure 5.6. For both kinds of transistors, the slanted line with the arrow represents the emitter, the slanted line with no arrow represents the collector, and the straight line*

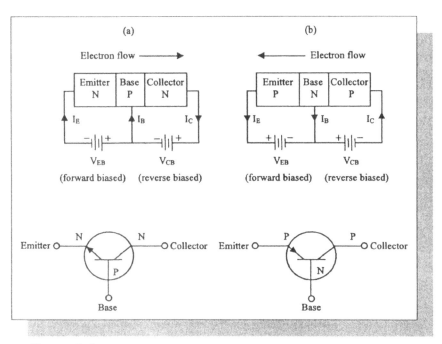

Figure 5.6
Junction transistor biasing diagrams and schematic symbols.

connecting the emitter and the collector represents the base. The
arrow always points in the direction of conventional current, which
is opposite to the direction of electron flow.

An easy way to identify the kind of transistor shown on a diagram is
to look at the emitter terminal:

● If the arrow points away from the base, the symbol represents an NPN
transistor.

● If the arrow points toward the base, the symbol represents a PNP
transistor.

5.3.1.1 THE NPN TRANSISTOR

For the NPN transistor in Figure 5.6(a), the **emitter** supplies electrons
to the base. Therefore, the symbol for the N emitter shows the arrow
out from the base, opposite to the direction of electron flow. Most NPN

transistors are made of silicon, with a typical forward bias of 0.6 V between base and emitter.

The **collector** (N) for the NPN transistor functions to receive electrons. The collector-base junction always has reverse bias (typically at values from 4 to 100 V). This polarity means that no majority charges can flow from collector to base. In the opposite direction, however, from base to collector, the collector voltage attracts the charges in the base supplied by the emitter.

The **base** (for both NPN and PNP) separates the emitter and collector. The base-emitter junction is forward-biased. As a result, the resistance for the emitter circuit is very low. The base-collector junction is reverse-biased, providing a much higher resistance in the collector circuit.

5.3.1.2 THE PNP TRANSISTOR

As shown for the PNP transistor in Figure 5.6(b), the P **emitter** supplies charges to its junction with the base. This direction is indicated by the emitter arrow for forward hole current in the schematic symbol.

The **collector** in a PNP transistor functions to remove charges from the junction with the base. In Figure 5.6(b), the PNP transistor has a P collector receiving hole charges.

As stated earlier, the **base** for the PNP and NPN separates the emitter and collector; it functions to control collector current.

In operation, a transistor carries several currents, including the emitter current (I_E), the base current (I_B), and the collector current (I_C). Because of the movement of the majority and minority carriers, changes in one current—for example, the emitter current—produce proportional changes in another current—for example, the collector current.

5.3.2 TRANSISTOR ACTION

The term, *transistor action*, originated by the inventors of the transistor, is used to describe the action of transferring current across a resistor—*trans* from transfer and *istor* from resistor. In the PNP transistor, transistor action occurs when the bias voltages pull holes from the emitter material through the base region. These holes move into the collector region, where the bias voltage continues to attract the holes through the material.

In the NPN transistor, transistor action is similar. However, this time, the emitter has a negative voltage, and the base has a positive voltage. The base region attracts free electrons from the emitter. These free electrons are moving fast enough to move right through the base. Once they enter the collector material, the positive bias voltage on the region continues to attract the electrons.

Note: *Transistor action should not be confused with transistor effect. Transistor effect occurs when current in a forward-biased*
IMPORTANT *diode is transferred across the resistance of a reverse-biased diode.*

5.4 TRANSISTOR CONNECTIONS

How a transistor is connected (especially in amplifiers) affects the way it performs. Because a transistor has only three electrodes, one must be a common connection for two pairs of terminals for input and output signals (see Figure 5.7). Specifically, in electronic circuits, the three possibilities for amplifier circuits are common base, common emitter, and common collector.

In a transistor with a **common base** connection, the input is to the emitter, and the output is from the collector as shown in Figure 5.8. The common base circuit is seldom used. It has no current gain from input to output, and although the voltage gain can be high, the output is shunted

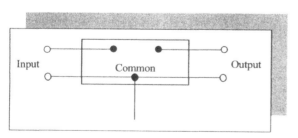

Figure 5.7
General case of three electrodes, with one common.

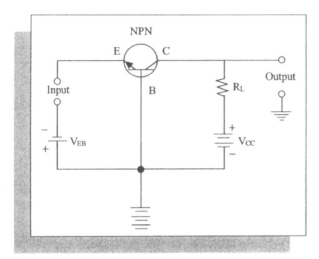

Figure 5.8
Common base
connection.

by the low resistance of the next stage. The only advantage of the common base circuit is that it has the best stability with an increase in temperature.

The most common transistor connection is the **common emitter** connection because it has the best combination of voltage gain and current gain. In this arrangement, the input is to the base, and the output is from the collector as shown in Figure 5.9.

The **common collector** connection usually is called an *emitter follower*, because the emitter voltage follows the base voltage very closely in response to a small input signal. The input for the emitter follower is to the base, and the output is from the emitter as shown in Figure 5.10.

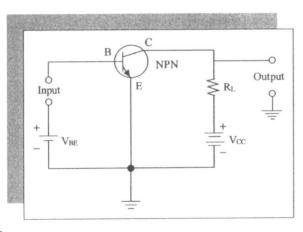

Figure 5.9
Common emitter
connection.

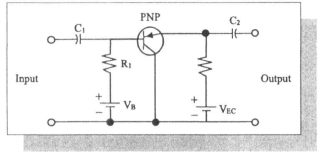

Figure 5.10
Common collector connection.

5.5 TRANSISTOR CHARACTERISTICS

As with all other electronic components, transistor performance can be described by certain parameters or characteristics. The characteristics important to transistor performance include

● **current gain**—depends partly on the kind of transistor connection

● **voltage gain**—the ratio of the output voltage to the input voltage

● **input resistance**—the ratio of the input voltage to the input current with the load connected

● **output resistance**—the ratio of the output voltage to the output current with the supply source connected

The values for these characteristics also vary depending on the kind of circuit connection. Table 5.1 compares characteristics for the different kinds of circuit connections.

TABLE 5.1. Transistor Characteristics (Based on Connection).			
Characteristic	Common-emitter	Common-collector	Common-base
Current gain	High	High	Low
Voltage gain	High	Low	High
Input resistance	Medium	High	Low
Output resistance	Medium-to-high	Low	High
Advantage	High gain	High input Rt	Stability

From Table 5.1, it should be obvious that no transistor connection method has ideal characteristics, but the common-emitter connection is used the most because of its high voltage gain and high current gain. The type of transistor connection used depends on the characteristics desired. For example, the common-emitter connection is desirable when the goal is to provide less source power to drive a load. The common-collector type might be appropriate where high input resistance is desired to prevent the device from drawing high power from the source. Moreover, the common-collector's low output resistance characteristic means that most of the output power is delivered to the load.

5.6 FIELD-EFFECT TRANSISTORS (FET)

To this point, the only transistor type described was the bipolar junction transistor (BJT). BJTs rely on an input-signal **current** to control the output current. The input-output current variations produce a varying output current. That output current is an amplified version of the input signal.

Another transistor type that has come into use is the *field effect transistor* (FET). The FET, like the BJT, is used in many switching and amplification applications. The FET is preferred when a high input impedance circuit is needed. The BJT has a relatively low input impedance as compared to the FET. (Note: The point is that the FET can take several volts for the input circuit, compared with tenths of a volt for junction transistors). Like the BJT, the FET is a three-terminal device. The terminals are called the **source, drain**, and **gate**. They are similar to the emitter, collector, and base, respectively.

 Key Point: *Field-effect transistors are considered voltage-driven devices, whereas bipolar transistors are current-driven.*

IMPORTANT

The basic design of an FET is shown in Figure 5.11. From the figure, it can be seen that the operation depends on controlling current through a semiconductor channel of one polarity. An N channel is shown here, but

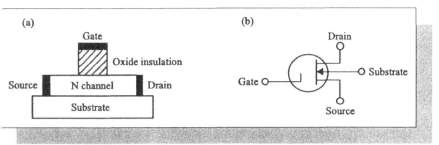

Figure 5.11
(a) Construction of FET; (b) schematic symbol for N-channel FET.

a P channel can be used instead. [Note: If a channel is of N material, it is called an N-channel FET (i.e., the arrow for hole charges into the channel indicates an N channel). If it is of P material, it is called a P-channel FET]. The substrate material is neutral or lightly doped silicon. This only serves as a platform on which the other electrodes are diffused.

When voltage is applied between the drain and the source at opposite ends of the channel, the gate controls the current through the channel.

The FET is a unipolar device, as the charge carriers in the channel have only one polarity. It also has a very high resistance from gate to source. The FET is also less sensitive to temperature. The disadvantages are less gain for a given bandwidth and smaller power ratings. Moreover, the switching speed is slower, compared with bipolar transistors.

5.7 THE JUNCTION FIELD EFFECT TRANSISTOR (JFET)

In the last section, we learned about **field-effect transistors**, or **FETs**. In the FET, an insulating layer is used in Figure 5.11 to isolate the gate from the N channel in the FET. However, the *junction field-effect transistor* (JFET) uses a different method of creating this isolation. The JFET uses a reverse-biased junction between the gate and the channel.

A cross section of an N-channel JFET is shown in Figure 5.12. The gate and N channel form a PN junction. The JFET is a depletion-mode

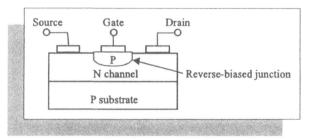

Figure 5.12
Cross section of N-channel JFET.

device in that the source-to-drain resistance is low until a negative voltage (reverse bias) is placed on the gate terminal. The negative voltage will **deplete** the current carriers in the N channel and shut off the current.

A schematic symbol for an N-channel JFET is shown in Figure 5.13. The channel here is N-type because the arrow is pointing toward the channel.

In operation, the N-channel JFET has a positive voltage applied to the drain with respect to the source. This allows a current to flow through the channel. If the gate is at 0 V, the drain current will be at its largest value for safe operation, and the JFET will be in the *on* condition. When a negative voltage is applied to the gate, the drain current will be reduced. As the gate voltage becomes more negative, the current decreases until cutoff, which occurs when the JFET is in the *off* condition.

Important Point: *The JFET is a normally on device, but the BJT is considered a normally off device. The JFET, therefore, can be*
IMPORTANT *utilized as a switching device.*

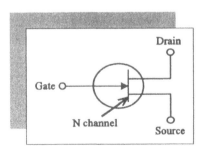

Figure 5.13
Schematic symbol for the N-channel JFET.

5.8 THE TRANSISTOR AS A SWITCH

We cannot ignore transistors in our daily lives; they are everywhere. This section introduces their most simple and widespread application—switching, with emphasis on the BJT.

In Section 5.7, we saw how the transistor can be turned *on* and *off* in a manner similar to a mechanical switch. Figure 5.14 shows two circuits (a) and (b), one of a transistor turned *on* and one of a transistor turned *off*. The transistor is switched by reversing the polarity of the input. For example, in circuit (a), the collector-to-ground output voltage is low and the circuit is said to be *on* because a collector current is present. In circuit (b), the polarity of the battery E_{B1} is reversed, which reduces the base current to about zero. The base-to-emitter section of the transistor is reverse-biased. Because the base current is almost zero, the collector current is about zero. Therefore, the collector-to-ground voltage is almost E_{B2} voltage. Because no current is present, this is equivalent to an open switch, and the circuit is said to be *off*.

The transistor in Figure 5.14(a) is said to be at **saturation**. It has reached its minimum resistance from emitter to collector, which produces the maximum collector current.

The transistor in Figure 5.14(b) is said to be at **cutoff**. It has reached its maximum resistance from emitter to collector and has cut off the current. The very tiny current still flowing is due to minority current carriers in the transistor, which is leakage current. When used as a switch, the transistor is driven to cutoff or to saturation by the emitter-base voltage.

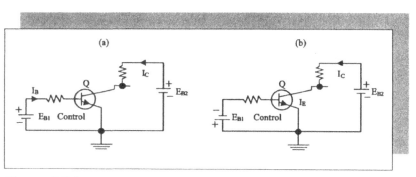

Figure 5.14
(a) Transistor on; (b) transistor off.

In summary, a small current in the control circuit (as designated in Figure 5.14) causes a large current to flow in the power circuit. With no current in the control circuit, the transistor acts like an open switch. With some current in the control circuit, the transistor acts like a closed switch.

5.9 TRANSISTOR SPECIFICATIONS

All electronic components have maximum voltage and current ratings. Transistors are no different. Transistors have several voltage and current ratings with which electronics technicians and/or maintenance operators responsible for electronic equipment or systems must be familiar. There are also several other specifications that are important to the technician. These specifications include electrical characteristics, physical characteristics, case styles, and power ratings, to name a few.

Fortunately, manufacturers typically provide data sheets or data books that list the specifications for each transistor. These specification sheets are extremely important to those who set up new systems. For normal maintenance and repair work, the original transistor usually is replaced with a duplicate.

5.10 INTEGRATED CIRCUITS

Technological advances in manufacturing techniques have ushered in an era of miniaturization in electronics. Nowhere can this profound influence of miniaturization be seen more than in the incorporation of transistors, thousands, or even millions, of them fabricated into a single device. These devices are extremely small and are called *integrated circuits* (ICs). ICs provide a complete circuit in one semiconductor package. They form the basis of the operation of many of our new products and are applied practically everywhere. Home computers are made possible because of this ability to miniaturize transistor-switching circuits.

Self-Test

5.1 What is meant by **current gain?**

5.2 In the JFET, what controls the flow of current?

5.3 Compared with bipolar transistors, FETs have a _____ input resistance.

5.4 The emitter-base junction of a bipolar transistor is always _____ biased.

5.5 The FET is a _____ driven device with high input impedance.

5.6 The base on a junction transistor serves the same purpose as the _____ on the FET.

5.7 An FET that is normally on is said to be a(n) _____-mode device.

5.8 A transistor with the common at the emitter has a _____ voltage gain and a _____ current gain.

5.9 A transistor operating in the saturation region acts as a switch that is turned _____.

5.10 A bipolar junction transistor is controlled by _____.

Alternating Current (A-C) Review

TOPICS

Generating A-C
Characteristic Values of A-C Voltage and Current
Resistance in A-C Circuits
Inductance
Capacitance
Inductive and Capacitive Reactance
Basic A-C Circuit Theory
Power in A-C Circuits

6.1 INTRODUCTION[1]

To this point, we have studied steady, unchanging voltages and currents (i.e., d-c voltage and current). We learned to use electronics principles to calculate circuit conditions with these steady-state voltages applied (i.e., battery supplied). Now, we are ready to consider the effects of "changing" voltages and currents.

A-c electronics is basically the study of the effects various circuit components have on sine-wave signals. Sine waves are a perfectly natural occurrence and are found in many places. Consider, for instance, the most common sine wave is the house current provided at a standard wall plug. Typically, this is a 120-volt sine wave that changes at a rate of 60 cycles per second, or at a frequency of 60 Hz (Hertz). This current

---

[1]Much of the information contained in this chapter is taken in condensed form, including many of the illustrations, from Spellman, F. R. and Drinan, J. *Fundamentals for the Water and Wastewater Maintenance Operators Series: Electricity.* Lancaster, PA: Technomic Publishing Co. Inc., 2000.

Key Terms Used in This Chapter

ALTERNATOR	An a-c generator.
ALTERNATING CURRENT AND VOLTAGE	Current and voltage that reverses between positive and negative polarities and varies in amplitude with time.
ALTERNATION	Variation, either positive or negative, of a waveform from zero to maximum and back to zero, equaling one-half a cycle.
AVERAGE VALUE	In sine-wave a-c voltage or current, 0.637 of peak value.
CYCLE	One complete set of values for a repetitive waveform.
EFFECTIVE VALUE	For sine-wave a-c waveform, 0.707 of peak value. Corresponds to heating effect of same d-c value. Also called rms value.
FREQUENCY	Number of recurrences of a periodic signal in a given time period, measured in hertz.
PEAK VALUE	Maximum amplitude, in either polarity; 1.414 times rms value for sine-wave E or I.
PHASE ANGLE	In an a-c circuit, the angle of lead or lag between the current and voltage waveforms. It is also the angle of the impedance of the circuit.
RMS VALUE	For sine-wave a-c waveform, 0.707 of peak value. Also called effective value.
SINE WAVE	One in which amplitudes vary in proportion to the sine function of an angle.
WAVELENGTH	Distance in space between two points with the same magnitude and direction in a propagated wave.

HYSTERESIS	In electromagnets, the effect of magnetic induction lagging in time behind the applied magnetizing force.
MUTUAL INDUCTION	Ability in one coil to induce voltage in another coil.
SELF-INDUCTANCE	The inductance produced in a coil by current in the coil itself.
INDUCTIVE REACTANCE	The opposition inductors have to current changes.
IMPEDANCE	Total opposition (resistance and reactance) by a circuit to a-c current flow, measured in ohms.
DIELECTRIC CONSTANT	Insulating material; it cannot conduct current but does store charge.
INDUCTANCE	Property of a circuit that opposes change in an existing current.
CAPACITANCE	The ratio between the electric charge transferred from one electrode of a capacitor to another and the resultant difference in potential between the electrodes.
CAPACITIVE REACTANCE	The opposition capacitors have to voltage changes.
APPARENT POWER	Product of $E \times I$ when they are out of phase. Measured in voltampere units, instead of watts.
POWER FACTOR	Cosine of the phase angle for a sine-wave a-c circuit. Value is between 1 and 0.
TRUE POWER	The net power consumed by resistance. Measured watts.
VECTOR QUANTITY	One that has magnitude and direction.

is produced and supplied by an electric utility system by a generator (or alternator).

Another example of sine waves is the sound from musical instruments. Moreover, many other electronic signals—such as speech—are complex combinations of many sine waves, all at different frequencies.

As stated above, the study of a-c electronics starts with the properties of simple sine waves and continues with variations in sine wave currents caused by differing voltages. In electronics, many sine wave voltages and currents are produced by circuits called oscillators, and these signals may change many millions of times a second. (Note: A detailed discussion of oscillators is covered in Chapter 9).

The purpose of this chapter is to provide maintenance operators with a review of the information required to understand basic a-c theory. Accordingly, we (1) explain how an alternator generates an a-c voltage; (2) explain frequency and state the factors that affect it; (3) analyze the sine waveform of a-c voltage or current; (4) make the mathematical conversions between maximum and effective values of voltage or current; (5) explain vectors, or phasors; (6) explain inductance; (7) explain capacitance; (8) explain reactance; (9) explain a-c circuit theory; and (10) explain power in a-c circuits.

The study of a-c starts with its generation, which is covered in the following section.

Key Point: *a-c electronics is a study of the effects various circuit components have on sine-wave signals.*

IMPORTANT

6.2 GENERATING A-C

Because voltage is induced in a conductor when lines of force are cut, the amount of the induced emf depends on the number of lines cut in a unit time. To induce an emf of 1 volt, a conductor must cut 100 million lines of force per second. To obtain this great number of cuttings, the conductor is formed into a loop and rotated on an axis at great speed (see Figure 6.1). The two sides of the loop become individual conductors in series, each side of the loop cutting lines of force and inducing twice the

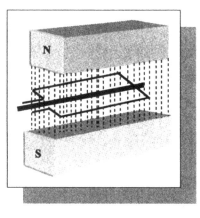

Figure 6.1
Loop rotating in magnetic field produces an a-c voltage.

voltage that a single conductor would induce. In commercial generators, the number of cuttings and the resulting emf are increased by (1) increasing the number of lines of force by using more magnets or stronger electromagnets, (2) using more conductors or loops, and (3) rotating the loops faster.

Unlike d-c voltage, a-c voltage can be stepped up or down by a device called a **transformer** (see Chapter 11). Transformers permit the transmission lines to be operated at high voltage and low current for maximum efficiency. Then, at the consumer end, the voltage is stepped down to whatever value the load requires by using a transformer.

6 2.1 BASIC A-C GENERATOR

An a-c voltage and current can be produced when a conductor loop rotates through a magnetic field and cuts lines of force to generate an induced a-c voltage across its terminals. This action describes the basic principle of operation of an alternating current generator, or **alternator**. An alternator converts mechanical energy into electrical energy. It does this by utilizing the principle of **electromagnetic induction**. The basic components of an alternator are an armature, about which many turns of conductor are wound, which rotates in a magnetic field, and some means of delivering the resulting alternating current to an external circuit.

Regarding electronics applications, we say that the a-c voltage source is usually produced by an alternator, which produces a regular output

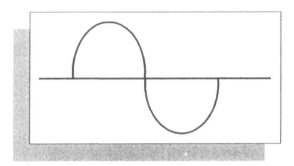

Figure 6.2
One cycle of a sine wave.

waveform, such as the *sine wave* (see Figure 6.2). In the laboratory, there are a number of electronic instruments that are used to produce sine waves. Sine-wave generators are quite useful because they allow us to adjust the voltage and frequency by turning a dial or pushing a button. The *function generator* is probably the most popular instrument used to produce the sine wave. It actually provides a choice of functions (or waveforms).

For our study, the term generator will simply mean a sine-wave source. Figure 6.3 shows the schematic symbol we use for an a-c generator; that is, the sine wave shown within the circle designates that it is an a-c sine-wave source.

To gain a better understanding of the sine wave, we need to study the cycle, which is covered in the following section.

6.2.1.1 CYCLE

An a-c voltage is one that continually changes in magnitude and periodically reverses in polarity (see Figure 6.4). The zero axis is a horizontal

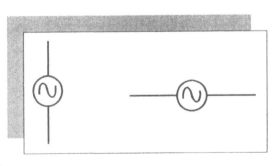

Figure 6.3
Schematic symbols used for an a-c source.

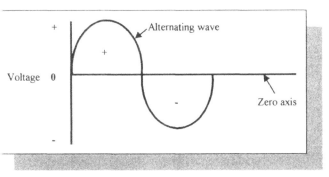

Figure 6.4
An a-c voltage waveform.

ine across the center. The vertical variations on the voltage wave show
he changes in magnitude. The voltages above the horizontal axis have
ositive (+) polarity, while voltages below the horizontal axis have
iegative (−) polarity. **Two alternations represent one complete circle
)f rotation.**

Important Point: *Two complete alternations in a period of time
is called a cycle—measured in the number of cycles per second
(cps), or* **hertz.**

Important Point: *One complete sine wave is called a* **cycle.**

6.2.1.2 FREQUENCY, PERIOD,
AND WAVELENGTH

The *frequency* of an alternating voltage or current is the number
)f complete cycles occurring in each second of time. It is indicated by
he symbol *f* and is expressed in hertz (Hz). One cycle per second equals
I Hz. Thus, 60 cycles per second (cps) equals 60 Hz. A frequency of 2 Hz
Figure 6.5(b)] is twice the frequency of 1 Hz [Figure 6.5(a)].

The amount of time for the completion of one cycle is the *period.*
t is indicated by the symbol *T* for time and is expressed in seconds (s).
=requency and period are reciprocals of each other.

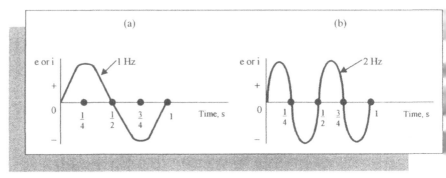

Figure 6.5
Comparison of frequencies.

$$f = \frac{1}{T} \qquad (6.1)$$

$$T = \frac{1}{f} \qquad (6.2)$$

 Important Point: The higher the frequency, the shorter the period.

IMPORTANT

 Important Point: Time measurements can be made from any point in the sine wave, but usually they are made from the point IMPORTANT at which it crosses the zero axis.

The angle of 360° represents the time for one cycle, or the period *T*. So we can show the horizontal axis of the sine wave in units of either electrical degrees or seconds (see Figure 6.6).

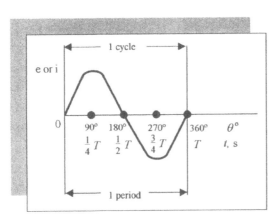

Figure 6.6
Relationship between electrical degrees and time.

90 *Alternating Current (A-C) Review*

Note: *The time to complete one sine wave is called the*
period (T).

The *wavelength* is the length of one complete wave or cycle. It depends upon the frequency of the periodic variation and its velocity of transmission. It is indicated by the symbol λ (Greek lowercase lambda). Expressed as a formula:

$$\lambda = \frac{\text{velocity}}{\text{frequency}} \qquad (6.3)$$

Important Point: *Frequency is related to time: f = 1/T.*

6.3 CHARACTERISTIC VALUES OF A-C VOLTAGE AND CURRENT

Because an a-c sine-wave voltage or current has many instantaneous values throughout the cycle, it is convenient to specify magnitudes for comparing one wave with another. The peak, average, or root mean square (rms) value can be specified (see Figure 6.7). These values apply to current or voltage.

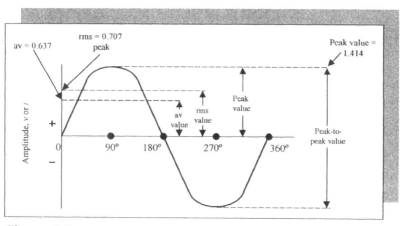

Figure 6.7
Amplitude values for a-c sine wave.

Important Point: *The most important voltage or amplitude measurements are the* **peak, peak-to-peak, average value,** *and* IMPORTANT *the* **root mean square (rms)** *voltages.*

6.3.1 PEAK AMPLITUDE

One of the most frequently measured characteristics of a sine wave is its amplitude. Unlike d-c measurement, the amount of alternating current or voltage present in a circuit can be measured in various ways. In one method of measurement, the maximum amplitude of either the positive or the negative alternation is measured. The value of current or voltage obtained is called the *peak voltage* or the *peak current*. To measure the peak value of current or voltage, an oscilloscope must be used. The peak value is illustrated in Figure 6.7.

6.3.2 PEAK-TO-PEAK AMPLITUDE

A second method of indicating the amplitude of a sine wave consists of determining the total voltage or current between the positive and negative peaks. This value of current or voltage is called the *peak-to-peak value* (see Figure 6.7). Because both alternations of a pure sine wave are identical, the peak-to-peak value is twice the peak value. Peak-to-peak voltage is usually measured with an oscilloscope, although some voltmeters have a special scale calibrated in peak-to-peak volts.

6.3.3 INSTANTANEOUS AMPLITUDE

The *instantaneous value* of a sine wave of voltage for any angle of rotation is expressed by the formula:

$$e = E_m \times \sin \theta \qquad (6.4)$$

where

e = the instantaneous voltage
E_m = the maximum or peak voltage
$\sin \theta$ = the sine of angle at which e is desired

Similarly, the equation for the instantaneous value of a sine wave of current would be

$$i = I_m \times \sin \theta \qquad (6.5)$$

where

i = the instantaneous current
I_m = the maximum or peak current
$\sin \theta$ = the sine of the angle at which i desired

Note: *The instantaneous value of voltage constantly changes as the armature of an alternator moves through a complete rotation.*
IMPORTANT *Because current varies directly with voltage, according to Ohm's Law, the instantaneous changes in current also result in a sine wave whose positive and negative peaks and intermediate values can be plotted exactly as we plotted the voltage sine wave. However, instantaneous values are not useful in solving most a-c problems, so an **effective** value is used.*

6.3.4 EFFECTIVE OR RMS VALUE

The *effective value* of an a-c voltage or current of sine waveform is defined in terms of an equivalent heating effect of a direct current. Heating effect is independent of the direction of current flow.

Important Point: *Because all instantaneous values of induced voltage are somewhere between zero and E_M (maximum, or peak*
IMPORTANT *voltage), the effective value of a sine-wave voltage or current must be greater than zero and less than E_M (maximum, or peak voltage).*

The alternating current of sine waveform having a maximum value of 14.14 amps produces the same amount of heat in a circuit having a resistance of 1 ohm as a direct current of 10 amps. Because this is true, we can work out a constant value for converting any peak value to a corresponding effective value. X represents this constant in the simple equation below. Solve for X to three decimal places.

$$14.14X = 10$$
$$X = 0.707$$

The effective value is also called the *root-mean-square (rms)* value because it is the square root of the average of the squared values between zero and maximum. The effective value of an a-c current is stated in terms of an equivalent d-c current. The phenomenon used as the standard comparison is the heating effect of the current.

 Important Point: *Any time an a-c voltage or current is stated without any qualifications, it is assumed to be an effective value.*

IMPORTANT

In many instances, it is necessary to convert from effective to peak or vice versa using a standard equation. Figure 6.7 shows that the peak value of a sine wave is 1.414 times the effective value; therefore, the equation we use is

$$E_m = E \times 1.414 \qquad (6.6)$$

where

E_m = maximum or peak voltage
E = effective or rms voltage

and

$$I_m = I \times 1.414 \qquad (6.7)$$

where

I_m = maximum or peak current
I = effective or rms current

Upon occasion, it is necessary to convert a peak value of current or voltage to an effective value. This is accomplished by using the following equations:

$$E = E_m \times 0.707 \qquad (6.8)$$

where

E = effective voltage
E_m = the maximum or peak voltage

$$I = I_m \times 0.707 \qquad (6.9)$$

where

I = the effective current
I_m = the maximum or peak current

6.3.5 AVERAGE VALUE

Because the positive alternation is identical to the negative alternation, the *average value* of a complete cycle of a sine wave is zero. In certain types of circuits, however, it is necessary to compute the average value of one alternation. Figure 6.7 shows that the average value of a sine wave is $0.637 \times$ peak value and, therefore,

$$\textbf{Average Value} = \textbf{0.637} \times \textbf{peak value} \qquad (6.10)$$

or

$$E_{avg} = E_m \times 0.637$$

where

E_{avg} = the average voltage of one alternation
E_m = the maximum or peak voltage

similarly

$$I_{avg} = I_m \times 0.637 \qquad (6.11)$$

where

I_{avg} = the average current in one alternation
I_m = the maximum or peak current

Table 6.1 lists the various values of sine-wave amplitude used to multiply in the conversion of a-c sine-wave voltage and current.

TABLE 6.1. A-C Sine-Wave Conversion Table.		
Multiply the Value	by	To Get the Value
Peak	2	Peak-to-peak
Peak-to-peak	0.5	Peak
Peak	0.637	Average
Average	1.637	Peak
Peak	0.707	rms (effective)
rms (effective)	1.414	Peak
Average	1.110	rms (effective)
rms (effective)	0.901	Average

6.4 RESISTANCE IN A-C CIRCUITS

The a-c signals are passed through resistors, just as d-c current is; that is, resistors interact with a-c signals just as they do with d-c currents.

If a sine wave of voltage is applied to a resistance, the resulting current will also be a sine wave. This follows Ohm's Law that states that the current is directly proportional to the applied voltage. Figure 6.8 shows a sine wave of voltage and the resulting sine wave of current superimposed on the same time axis. Notice that as the voltage increases in a positive direction, the current increases along with it. When the voltage reverses direction, the current reverses direction. At all times, the voltage and current pass through the same relative parts of their respective cycles at the same time. When two waves, such as those shown in Figure 6.8, are precisely in step with one another, they are said to be **in phase**. To

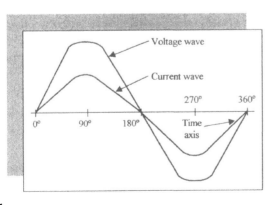

Figure 6.8
Voltage and current waves in phase.

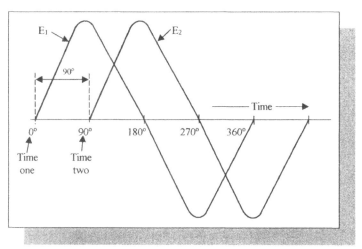

Figure 6.9
Voltage waves 90° out of phase.

be in phase, the two waves must go through their maximum and minimum points at the same time and in the same direction.

In some circuits, several sine waves can be in phase with each other. Thus, it is possible to have two or more voltage drops in phase with each other and also in phase with the circuit current.

IMPORTANT

Note: *It is important to remember that Ohm's Law for d-c circuits is applicable to a-c circuits* **with resistance only.**

Voltage waves are not always in phase. For example, Figure 6.9 shows a voltage wave E_1 considered to start at 0° (time 1). As voltage wave E_1 reaches its positive peak, a second voltage wave E_2 starts to rise (time 2). Because these waves do not go through their maximum and minimum points at the same instant of time, a **phase difference** exists between the two waves. The two waves are said to be out of phase. For the two waves in Figure 6.9, this phase difference is 90°.

6.4.1 PHASE RELATIONSHIPS

In the preceding section, we discussed the important concepts of **in phase** and **phase difference**. Another important phase concept is phase

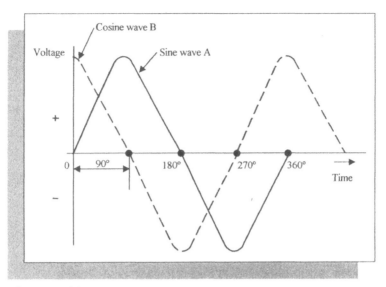

Figure 6.10
Wave B leads wave A by a phase angle of 90°.

angle. The *phase angle* between two waveforms of the same frequency is the angular difference at a given instant of time. As an example, the phase angle between waves B and A (see Figure 6.10) is 90°. Take the instant of time at 90°. The horizontal axis is shown in angular units of time. Wave B starts at maximum value and reduces to zero value at 90°, while wave A starts at zero and increases to maximum value at 90°. Wave B reaches its maximum value 90° ahead of wave A, so wave B **leads** wave A by 90° (and wave A **lags** wave B by 90°). This 90° phase angle between waves B and A is maintained throughout the complete cycle and all successive cycles. At any instant of time, wave B has the value that wave A will have 90° later. Wave B is a cosine wave because it is displaced 90° from wave A, which is a sine wave.

 Important Point: *The amount by which one wave leads or lags another is measured in degrees.*

IMPORTANT

To compare phase angles or phases of alternating voltages or currents, it is more convenient to use vector diagrams corresponding to the voltage and current waveforms. A *vector* is a straight line used to denote the magnitude and direction of a given quantity. The length of the line, drawn

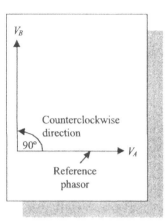

Figure 6.11
Phasor diagram.

to scale, denotes magnitude, and the direction is indicated by the arrow
at one end of the line, together with the angle that the vector makes with
a horizontal reference vector.

Note: *In electricity, because different directions really represent*
time expressed as a phase relationship, an electrical vector is called
IMPORTANT *a phasor. In an a-c circuit containing only resistance, the voltage*
and current occur at the same time, or are in phase. To indicate
this condition by means of phasors, all that is necessary is to draw
the phasors for the voltage and current in the same direction. The
length of the phasor indicates the value of each.

A vector, or phasor, diagram is shown in Figure 6.11 where vector V_B
is vertical to show the phase angle of 90° with respect to vector V_A, which
is the reference. Because lead angles are shown in the counterclockwise
direction from the reference vector, V_B leads V_A by 90°.

6.5 INDUCTANCE

Before discussing the inductor in an a-c circuit, it is important to
review key points about magnetic fields:

● A field of force exists around a wire carrying a current.

● This field has the form of concentric circles around the wire, in planes
perpendicular to the wire, and with the wire at the center of the circles.

● The strength of the field depends on the current. Large currents produce large fields; small currents produce small fields.

● When lines of force cut across a conductor, a voltage is induced in the conductor.

To this point, we have studied **resistive** circuits (i.e., resistors presented the only opposition to current flow). Two other phenomena—inductance and capacitance—exist in d-c circuits to some extent, but they are major players in a-c circuits. Both inductance and capacitance present a kind of opposition to current flow that is called **reactance** (covered in Section 6.7). Before we examine reactance, however, we must first study inductance and, later, capacitance.

Simply put, an *inductor* is a coil of wire, usually many turns around a piece of soft iron. In some cases, the wire is wound around a nonconducting material. The property of *inductance* is the characteristic of an electrical circuit that makes itself evident by opposing the starting, stopping, or changing of current flow. A simple analogy can be used to explain inductance. We are all familiar with how difficult it is to push a heavy load (a cart full of heavy materials, etc.). It takes more work to start the load moving than it does to keep it moving. This is because the load possesses the property of **inertia**. Inertia is the characteristic of mass that opposes a **change** in velocity. Therefore, inertia can hinder us in some ways and help us in others. Inductance exhibits the same effect on current in an electric circuit as inertia does on velocity of a mechanical object. The effects of inductance are sometimes desirable, and sometimes undesirable.

 Important Point: *Simply put, **inductance** is the characteristic of an electrical conductor that opposes a **change** in current flow.*

IMPORTANT

This means that, because inductance is the property of an electric circuit that opposes any **change** in the current through that circuit, if the current increases, a self-induced voltage opposes this change and delays the increase. On the other hand, if the current decreases, a self-induced voltage tends to aid or prolong the current flow, delaying the decrease. Thus, current can neither increase nor decrease as fast in an inductive circuit as it can in a purely resistive circuit.

In a-c circuits, this effect becomes very important because it affects the **phase** relationships between voltage and current. We learned earlier that voltages (or currents) can be out of phase if they are induced in separate armatures of an alternator. In that case, the voltage and current generated by each armature were in phase. When inductance is a factor in a circuit, the voltage and current generated by the **same** armature are out of phase. We shall examine these phase relationships later in this manual.

6.5.1 UNIT OF INDUCTANCE

The unit for measuring inductance, L, is the *henry* (named for the American physicist, Joseph Henry), abbreviated h. Figure 6.12 shows the schematic symbol for an inductor. An inductor has an inductance of 1 henry if an emf of 1 volt is induced in the inductor when the current through the inductor is changing at the rate of 1 ampere per second. The relation between the induced voltage, inductance, and rate of change of current with respect to time is stated mathematically as

$$E = L\frac{\Delta I}{\Delta t} \qquad (6.12)$$

where

E = the induced emf in volts
L = the inductance in henrys
ΔI = is the change in amperes occurring in Δt seconds

Note: *The symbol* Δ *(delta) means a change in.*

IMPORTANT

The henry is a large unit of inductance and is used with relatively large inductors. The unit employed with small inductors is the millihenry (mh). For still smaller inductors, the unit of inductance is the microhenry (μh).

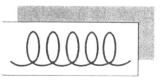

Figure 6.12
Schematic symbol for an inductor.

6.5.2 SELF-INDUCTANCE

As previously explained, current flow in a conductor always produces a magnetic field surrounding, or linking with, the conductor. When the current changes, the magnetic field changes, and an emf is induced in the conductor. This emf is called a *self-induced emf* because it is induced in the conductor carrying the current.

 Note: *Even a perfectly straight length of conductor has some inductance.*

IMPORTANT

The direction of the induced emf has a definite relation to the direction in which the field that induces the emf varies. When the current in a circuit is increasing, the flux linking with the circuit is increasing. This flux cuts across the conductor and induces an emf in the conductor in such a direction to oppose the increase in current and flux. This emf is sometimes referred to as **counterelectromotive force** (cemf). The two terms are used synonymously throughout this manual. Likewise, when the current is decreasing, an emf is induced in the opposite direction and opposes the decrease in current.

 Important Point: *The effects just described are summarized by Lenz's Law, which states that the induced emf in any circuit is* IMPORTANT *always in a direction opposed to the effect that produced it.*

Shaping a conductor so the electromagnetic field around each portion of the conductor cuts across another portion of the same conductor increases the inductance. This is shown in its simplest form in Figure 6.13(a). A loop of conductor is looped so two portions of the conductor lie adjacent and parallel to one another. These portions are labeled conductor 1 and conductor 2. When the switch is closed, electron flow through the conductor establishes a typical concentric field around **all** portions of the conductor. The field is shown in a single plane (for simplicity) that is perpendicular to both conductors. Although the field originates simultaneously in both conductors, it is considered as originating in conductor 1, and its effect on conductor 2 will be noted. With increasing current, the field expands outward, cutting across a portion of conductor 2. The dashed

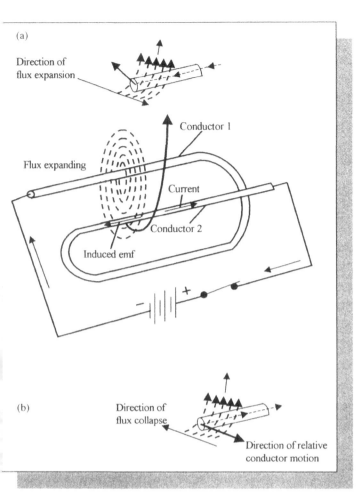

Figure 6.13
Self-inductance.

arrow shows the resultant induced emf in conductor 2. Note that it is in **opposition** to the battery current and voltage, according to Lenz's Law.

In Figure 6.13(b), the same section of conductor 2 is shown, but with the switch opened and the flux collapsing.

Important Point: *From Figure 6.13, the important point to note is that the voltage of self-induction opposes both changes in current. It delays the initial buildup of current by opposing the battery voltage and delays the breakdown of current by exerting an induced voltage in the same direction that the battery voltage acted.*

IMPORTANT

Figure 6.14
(a) Few turns, low inductance; (b) more turns, higher inductance.

Four major factors affect the self-inductance of a conductor, or circuit.

1. Number of turns: Inductance depends on the number of wire turns. Wind more turns to increase inductance. Take turns off to decrease the inductance. Figure 6.14 compares the inductance of two coils made with different numbers of turns.

2. Spacing between turns: Inductance depends on the spacing between turns, or the inductor's length. Figure 6.15 shows two inductors with the same number of turns. The first inductor's turns have a wide spacing. This coil is 6 cm long. The second inductor's turns are close together. The second coil is only 1 cm long. The second coil, though shorter, has a larger inductance value because of its close spacing between turns.

3. Coil diameter: Coil diameter, or cross-sectional area, is highlighted in Figure 6.16. The larger-diameter inductor has more inductance. Both coils shown have the same number of turns, and the spacing between turns is the same. The first inductor has a small diameter, and the second one has a larger diameter. The second inductor has more inductance than the first one.

4. Type of core material: **Permeability**, as pointed out earlier, is a measure of how easily a magnetic field goes through a material. Permeability also tells us how much stronger the magnetic field will be with the material inside the coil.

Figure 6.15
(a) Wide spacing between turns, low inductance; (b) close spacing between turns, higher inductance.

Figure 6.16
(a) Small diameter, low inductance; (b) larger diameter, higher inductance.

Figure 6.17 shows three identical coils. One has an air core, one has a powdered-iron core in the center, and the other has a soft iron core. This figure illustrates the effects of core material on inductance. The inductance of a coil is affected by the magnitude of current when the core is a magnetic material. When the core is air, the inductance is independent of the current.

Key Point: *The inductance of a coil increases very rapidly as the number of turns is increased. It also increases as the coil is made* IMPORTANT *shorter, the cross-sectional is made larger, or the permeability of the core is increased.*

6.5.3 GROWTH AND DECAY OF CURRENT IN AN RL SERIES CIRCUIT

If a battery is connected across a pure inductance, the current builds up to its final value at a rate that is determined by the battery voltage and the internal resistance of the battery. The current buildup is gradual because of the counter emf (cemf) generated by the self-inductance of the

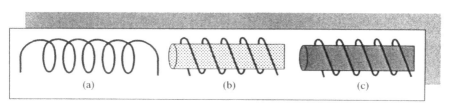

Figure 6.17
(a) Air core, low inductance; (b) powdered iron core, higher inductance; (c) soft iron core, highest inductance.

coil. When the current starts to flow, the magnetic lines of force move out, cut the turns of wire on the inductor, and build up a cemf that opposes the emf of the battery. This opposition causes a delay in the time it takes the current to build up to steady value. When the battery is disconnected, the lines of force collapse, again cutting the turns of the inductor and building up an emf that tends to prolong the current flow.

Although the analogy is not exact, electrical inductance is somewhat like mechanical inertia. A boat begins to move on the surface of water at the instant a constant force is applied to it. At this instant, its rate of change of speed (acceleration) is greatest, and all the applied force is used to overcome the inertia of the boat. After a while, the speed of the boat increases (its acceleration decreases), and the applied force is used up in overcoming the friction of the water against the hull. As the speed levels off and the acceleration becomes zero, the applied force equals the opposing friction force at this speed, and the inertia effect disappears. In the case of inductance, it is electrical inertia that must be overcome.

Note: *In many electronic circuits, the time required for the growth or decay of current is important. However, these applications are* IMPORTANT *beyond the scope of this manual, but you need to learn the "fundamentals" of the L/R time constant, which are discussed in the following section.*

6.5.4 *L/R* TIME CONSTANT

The time required for the current through an inductor to increase to 63.2% (63%) of the maximum current or to decrease to 36.7 % (37%) is known as the *time constant* of the circuit. An RL circuit is shown in Figure 6.18.

The value of the time constant in seconds is equal to the inductance in henrys divided by the circuit resistance in ohms. *L/R* is the symbol used for this time constant. If *L* is in henrys and *R* is in ohms, *t* (time) is in seconds. If *L* is in microhenrys and *R* is in ohms, *t* is in microseconds. If *L* is in millihenrys and *R* is in ohms, *t* is in milliseconds.

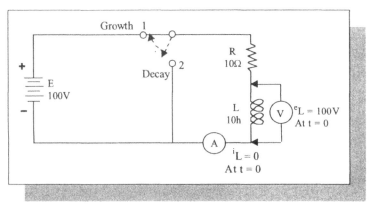

Figure 6.18
L/R *time constant.*

R in the L/R equation is always in ohms, and the time constant is on the same order of magnitude as L. Two useful relations used in calculating L/R time constants are as follows:

$$\frac{L \text{ (in henrys)}}{R \text{ (in ohms)}} = t \text{ (in seconds)} \qquad (6.13)$$

$$\frac{L \text{ (microhenrys)}}{R \text{ (in ohms)}} \qquad (6.14)$$

Key Point: *The time constant of an L/R circuit is always expressed as a ratio between inductance (or L) and resistance (or R).*

IMPORTANT

6.5.5 MUTUAL INDUCTANCE

When the current in a conductor or coil changes, the varying flux can cut across any other conductor or coil located nearby, thus inducing voltages in both (see Figure 6.19).

The factors affecting the mutual inductance of two adjacent coils is dependent upon

● physical dimensions of the two coils

● number of turns in each coil

● distance between the two coils

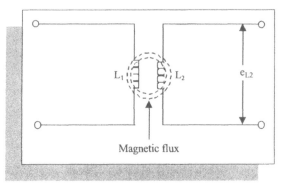

Figure 6.19
*Mutual inductance
between L_1 and I_2.*

Magnetic flux

- relative positions of the axes of the two coils

- the permeability of the cores

IMPORTANT

Important Point: *The amount of mutual inductance depends on the relative position of the two coils. If the coils are separated a considerable distance, the amount of flux common to both coils is small, and the mutual inductance is low. Conversely, if the coils are close together so that nearly all the flow of one coil links the turns of the other, mutual inductance is high. The mutual inductance can be increased greatly by mounting the coils on a common iron core.*

6.5.6 CALCULATION OF TOTAL INDUCTANCE

IMPORTANT

Note: *In the study of advanced electrical theory, it is necessary to know the effect of mutual inductance in solving for total inductance in both series and parallel circuits. However, for our purposes in this manual, we do not attempt to make these calculations. Instead, we discuss the basic total inductance calculations with which the maintenance operator should be familiar.*

If inductors in series are located far enough apart, or are well shielded to make the effects of mutual inductance negligible, the total inductance is calculated in the same manner as for resistances in series; we merely add them:

$$L_t = L_1 + L_2 + L_3 \dots \text{(etc.)} \qquad (6.15)$$

EXAMPLE 6.1

Problem: If a series circuit contains three inductors whose values are 40 μh, 50 μh, and 20 μh, what is the total inductance?

Solution:

$$L_t = 40 \text{ μh} + 50 \text{ μh} + 20 \text{ μ}$$
$$= 110 \text{ μh}$$

In a parallel circuit containing inductors (without mutual inductance), the total inductance is calculated in the same manner as for resistances in parallel:

$$\frac{1}{L_t} = \frac{1}{L_1} + \frac{1}{L_2} + \frac{1}{L_3} + \cdots \text{ (etc.)} \qquad (6.16)$$

EXAMPLE 6.2

Problem: A circuit contains three totally shielded inductors in parallel. The values of the three inductances are 4 mh, 5 mh, and 10 mh. What is the total inductance?

Solution:

$$\frac{1}{L_t} = \frac{1}{4} + \frac{1}{5} + \frac{1}{10}$$
$$= 0.25 + 0.2 + 0.1$$
$$= 0.55$$
$$L_t = \frac{1}{0.55}$$
$$= 1.8 \text{ mh}$$

6.6 CAPACITANCE

In Section 6.5, we learned that inductance opposes any change in current. *Capacitance* is the property of an electric circuit that opposes any change of *voltage* in a circuit. That is, if applied voltage is increased, capacitance opposes the change and delays the voltage increase across the circuit. If applied voltage is decreased, capacitance tends to maintain the higher original voltage across the circuit, thus delaying the decrease.

Capacitance is also defined as that property of a circuit that enables energy to be stored in an electric field. Natural capacitance exists in many electric circuits. However, in this manual, we are concerned only with the capacitance that is designed into the circuit by means of devices called **capacitors**.

 Key Point: *The most noticeable effect of capacitance in a circuit is that voltage can neither increase nor decrease as rapidly in a capacitive circuit as it can in a circuit that does not include capacitance.*

IMPORTANT

6.6.1 THE CAPACITOR

A *capacitor*, or condenser, is a manufactured electrical device that consists of two conducting plates of metal separated by an insulating material called a *dielectric* (see Figure 6.20). (Note: The prefix, di-means through or across.)

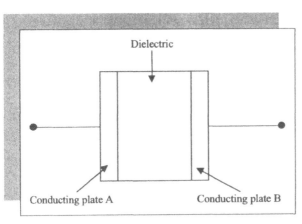

Figure 6.20
Capacitor.

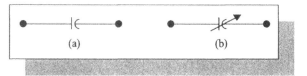

Figure 6.21
(a) Schematic for a fixed capacitor; (b) variable capacitor.

The schematic symbol for a capacitor is shown in Figure 6.21.

When a capacitor is connected to a voltage source, there is a short current pulse. A capacitor stores this electric charge in the dielectric (it can be charged and discharged, as we shall see later). To form a capacitor of any appreciable value, however, the area of the metal pieces must be quite large, and the thickness of the dielectric must be quite small.

Key Point: *A capacitor is essentially a device that stores electrical energy.*

IMPORTANT

The capacitor is used in a number of ways in electrical circuits. It may block d-c portions of a circuit, because it is effectively a barrier to direct current (but not to a-c current). It may be part of a tuned circuit—one such application is in the tuning of a radio to a particular station. It may be used to filter a-c out of a d-c circuit. Most of these are advanced applications that are beyond the scope of this manual; however, a basic understanding of capacitance is necessary to the fundamentals of a-c theory.

Important Point: *A capacitor does not conduct d-c current. The insulation between the capacitor plates blocks the flow*
IMPORTANT *of electrons. We learned earlier there is a short current pulse when we first connect the capacitor to a voltage source. The capacitor quickly charges to the supply voltage, and then the current stops.*

The two plates of the capacitor shown in Figure 6.22 are electrically neutral because there are as many protons (positive charge) as electrons (negative charge) on each plate. Thus, the capacitor has **no charge**.

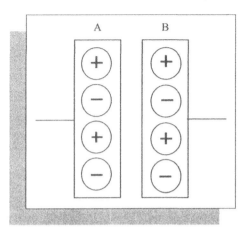

Figure 6.22
Two plates of a capacitor with a
neutral charge.

Now a battery is connected across the plates [see Figure 6.23(a)]. When the switch is closed [see Figure 6.23(b)], the negative charge on plate A is attracted to the positive terminal of the battery. This movement of charges will continue until the difference in charge between plates A and B is equal to the electromotive force (voltage) of the battery.

The capacitor is now **charged**. Because almost none of the charge can cross the space between plates, the capacitor will remain in this

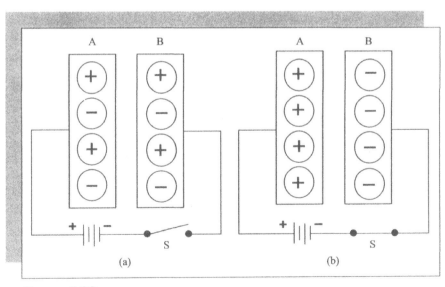

Figure 6.23
(a) Neutral capacitor; (b) charged capacitor.

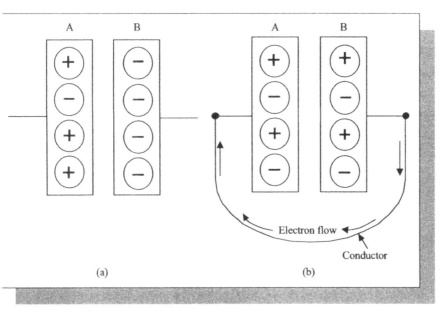

Figure 6.24
(a) Charged capacitor; (b) discharging a capacitor.

condition even if the battery is removed [see Figure 6.24(a)]. However, if a conductor is placed across the plates [see Figure 6.24(b)], the electrons find a path back to plate A, and the charges on each plate are again neutralized. The capacitor is now **discharged**.

 Important Point: *In a capacitor, electrons cannot flow through the dielectric, because it is an insulator. Because it takes a definite*
IMPORTANT *quantity of electrons to charge (fill up) a capacitor, it is said to have* ***capacity****. This characteristic is referred to as capacitance.*

6.6.2 DIELECTRIC MATERIALS

Somewhat similar to the phenomenon of permeability in magnetic circuits, various materials differ in their ability to support electric flux (lines of force) or to serve as dielectric material for capacitors. Materials are rated in their ability to support electric flux in terms of a number called a *dielectric constant*. Other factors being equal, the higher the value of

TABLE 6.2. Dielectric Constants	
Material	Constant
Vacuum	1.0000
Air	1.0006
Paraffin paper	3.5
Glass	5–10
Quartz	3.8
Mica	3–6
Rubber	2.5–35
Wood	2.5–8
Porcelain	5.1–5.9
Glycerin (15°C)	56
Petroleum	2
Pure water	81

the dielectric constant, the better is the dielectric material. Dry air is the standard (the reference) by which other materials are rated.

Dielectric constants for some common materials are given in Table 6.2.

Note: *From Table 6.2, it is obvious that pure water is the best dielectric. Keep in mind that the key word is pure. Water capacitors* IMPORTANT *are used today in some high-energy applications, in which differences in potential are measured in thousands of volts.*

6.6.3 UNIT OF CAPACITANCE

Capacitance is equal to the amount of charge that can be stored in a capacitor divided by the voltage applied across the plates:

$$C = \frac{Q}{E} \qquad (6.17)$$

where

C = capacitance, F (farads)
Q = amount of charge, C (coulombs)
E = voltage, V

EXAMPLE 6.3

Problem: What is the capacitance of two metal plates separated by 1 cm of air, if 0.002 coulomb of charge is stored when a potential of 300 volts is applied to the capacitor?

Solution:

Given:
$$Q = 0.002 \ coulomb$$
$$E = 300 \ volts$$
$$C = \frac{Q}{E}$$

Converting to power of 10

$$C = \frac{2 \times 10^{-3}}{3 \times 10^2}$$
$$C = 6.67 \times 10^{-6}$$
$$C = 0.00000667 \ farads$$

MPORTANT

Note: *Although the capacitance value obtained in Example 6.3 appears small, many electronic circuits require capacitors of much smaller value. Consequently, the farad is a cumbersome unit, far too large for most applications. The* **microfarad,** *which is one millionth of a farad (1 × 10⁻⁶ farad), is a more convenient unit. The symbols used to designate microfarad are μf.*

Equation (6.1) can be rewritten as follows:

$$Q = CE \qquad\qquad (6.18)$$

$$E = \frac{Q}{C} \qquad\qquad (6.19)$$

MPORTANT

Important Point: *From Equation (6.17), do not mistakenly assume that capacitance is dependent upon charge and voltage. Capacitance is determined entirely by physical factors, which are covered in Section 6.6.4.*

The symbol used to designate a capacitor is (C). The unit of capacitance is the farad (F). The farad is that capacitance that will store one coulomb of charge in the dielectric when the voltage applied across the capacitor terminals is one volt.

6.6.4 FACTORS AFFECTING THE VALUE OF CAPACITANCE

The capacitance of a capacitor depends on three main factors: plate surface area, distance between plates, and dielectric constant of the insulating material.

● **Plate surface area**—capacitance varies directly with place surface area. We can double the capacitance value by doubling the capacitor's plate surface area. Figure 6.25 shows a capacitor with a small surface area and another one with a large surface area.

Adding more capacitor plates can increase the plate surface area. Figure 6.26 shows alternate plates connecting to opposite capacitor terminals.

● **Distance between plates**—capacitance varies inversely with the distance between plate surfaces. The capacitance increases when the plates are closer together. Figure 6.27 shows capacitors with the same plate surface area, but with different spacing.

● **Dielectric constant of the insulating material**—an insulating material with a higher dielectric constant produces a higher capacitance

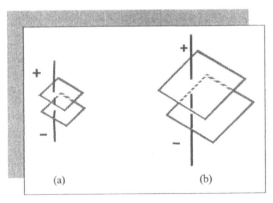

Figure 6.25
(a) Small plates, small capacitance; (b) larger plates, higher capacitance.

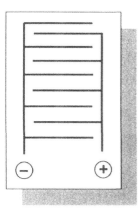

Figure 6.26
Shows several sets of plates connected to produce a capacitor with more surface area.

rating. Figure 6.28 shows two capacitors. Both have the same plate surface area and spacing. Air is the dielectric in the first capacitor, and mica is the dielectric in the second one. Mica's dielectric constant is 5.4 times greater than air's dielectric constant. The mica capacitor has 5.4 times more capacitance than the air-dielectric capacitor.

6.6.5 CHARGE AND DISCHARGE OF AN RC SERIES CIRCUIT

According to Ohm's Law, the voltage across a resistance is equal to the current through it times the value of the resistance. This means that a

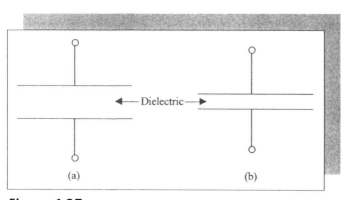

Dielectric

(a)   (b)

Figure 6.27
(a) Wide plate spacing, small capacitance; (b) narrow plate spacing, larger capacitance.

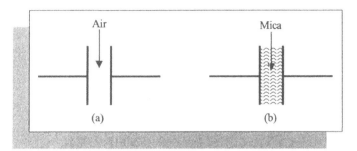

Figure 6.28
(a) Low capacitance; (b) higher capacitance.

voltage will be developed across a resistance **only when current flows through it**.

As previously stated, a capacitor is capable of storing or holding a charge of electrons. When uncharged, both plates contain the same number of free electrons. When charged, one plate contains more free electrons than the other does. The difference in the number of electrons is a measure of the charge on the capacitor. The accumulation of this charge builds up a voltage across the terminals of the capacitor, and the charge continues to increase until this voltage equals the applied voltage. The greater the voltage, the greater the charge on the capacitor. Unless a discharge path is provided, a capacitor keeps its charge indefinitely. Any practical capacitor, however, has some leakage through the dielectric so that the voltage will gradually leak off.

The actual time it takes to charge or discharge is important in advanced electricity and electronics. Because the charge or discharge time depends on the values of resistance and capacitance, an RC circuit can be designed for the proper timing of certain electrical events. *RC* time constant is covered in the next section.

6.6.6 *RC* TIME CONSTANT

The time required to charge a capacitor to 63% of maximum voltage or to discharge it to 37% of its final voltage is known as the *time constant* of the current. An RC circuit is shown in Figure 6.29.

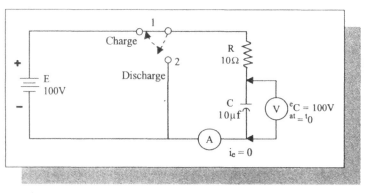

Figure 6.29
An RC circuit.

The time constant T for an RC circuit is

$$T = RC \qquad (6.20)$$

The time constant of an RC circuit is usually very short because the capacitance of a circuit may be only a few microfarads or even picofarads.

Key Point: *An RC time constant expresses the charge and discharge times for a capacitor.*

IMPORTANT

6.6.7 CAPACITORS IN SERIES AND PARALLEL

Like resistors or inductors, capacitors may be connected in series, in parallel, or in a series-parallel combination. Unlike resistors or inductors, however, total capacitance in series, in parallel, or in a series-parallel combination is found in a different manner. Simply put, the rules are not the same for the calculation of total capacitance. This difference is explained as follows:

Parallel capacitance is calculated like series resistance, and series capacitance is calculated like parallel resistance.

For example:

When capacitors are connected in **series** (see Figure 6.30), the total capacitance C_T is

Series: $$\frac{1}{C_T} = \frac{1}{C_1} + \frac{1}{C_2} + \frac{1}{C_3} + \cdots + \frac{1}{C_n} \qquad (6.21)$$

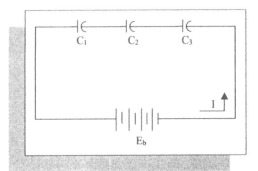

Figure 6.30
Series capacitive circuit.

EXAMPLE 6.4

Problem: Find the total capacitance of a 3-μF, a 5-μF, and a 15-μF capacitor in series.

Solution: Write Equation (6.21) for three capacitors in series.

$$\frac{1}{C_T} = \frac{1}{C_1} + \frac{1}{C_2} + \frac{1}{C_3}$$

$$= \frac{1}{3} + \frac{1}{5} + \frac{1}{15} = \frac{9}{15} = \frac{3}{5}$$

$$\frac{1}{C_T} = \frac{3}{5}$$

$$C_T = 1.7 \ \mu F$$

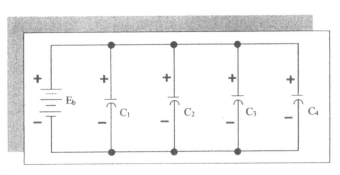

Figure 6.31
Parallel capacitive circuit.

When capacitors are connected in **parallel** (see Figure 6.31), the total capacitance C_T is the sum of the individual capacitances.

Parallel: $$C_T = C_1 + C_2 + C_3 + \cdots + C_n \qquad (6.22)$$

EXAMPLE 6.5

Problem: Determine the total capacitance in a parallel capacitive circuit:

Given: $C_1 = 2\ \mu F$
$C_2 = 3\ \mu F$
$C_3 = 0.25\ \mu F$

Solution: Write Equation (6.22) for three capacitors in parallel:

$$C_T = C_1 + C_2 + C_3$$
$$= 2 + 3 + 0.25$$
$$= 5.25\ \mu F$$

If capacitors are connected in a combination of **series and parallel** (see Figure 6.32), the total capacitance is found by applying Equations (6.21) and (6.22) to the individual branches.

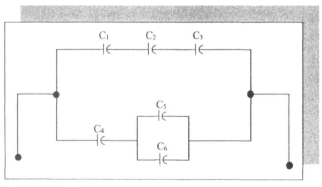

Figure 6.32
Series-parallel capacitance configuration.

TABLE 6.3. Comparison of Capacitor Types.		
Dielectric	Construction	Capacitance Range
Air	Meshed plates	10–400 pF
Mica	Stacked plates	10–5000 pF
Paper	Rolled foil	0.001–1 µF
Ceramic	Tubular disk	0.5–1600 pF
		0.002–0.1 µF
Electrolytic	Aluminum tantalum	5–1000 µF
		0.01–300 µF

6.6.8 TYPES OF CAPACITORS

Capacitors used for commercial applications are divided into two major groups—fixed and variable—and are named according to their dielectric. Most common are air, mica, paper, and ceramic capacitors, plus the electrolytic type. These types are compared in Table 6.3.

The fixed capacitor has a set value of capacitance that is determined by its construction. The construction of the variable capacitor allows a range of capacitances. Within this range, the desired value of capacitance is obtained by some mechanical means, such as by turning a shaft (as in turning a radio tuner knob, for example) or adjusting a screw to adjust the distance between the plates.

The electrolytic capacitor consists of two metal plates separated by an electrolyte. The electrolyte, either paste or liquid, is in contact with the negative terminal, and this combination forms the negative electrode. The dielectric is a very thin film of oxide deposited on the positive electrode, which is aluminum sheet. Electrolytic capacitors are polarity sensitive (i.e., they must be connected in a circuit according to their polarity markings) and are used where a large amount of capacitance is required.

6.7 INDUCTIVE AND CAPACITIVE REACTANCE

In Sections 6.5 and 6.6, we learned that the inductance of a circuit acts to oppose any change of current flow in that circuit and that capacitance acts to oppose any change of voltage. In d-c circuits, these **reactions** are not important, because they are momentary and occur only when a circuit is first closed or opened. In a-c circuits, these effects become very important, because the direction of current flow is reversed many times each second; and the opposition presented by inductance and capacitance is, for practical purposes, constant.

In purely resistive circuits, either d-c or a-c, the term for opposition to current flow is resistance. When the effects of capacitance or inductance are present, as they often are in a-c circuits, the opposition to current flow is called *reactance*. The total opposition to current flow in circuits that have both resistance and reactance is called *impedance*.

In this section, we cover the calculation of inductive and capacitive reactance and impedance; the phase relationships of resistance, inductive, and capacitive circuits; and power in reactive circuits.

6.7.1 INDUCTIVE REACTANCE

To gain understanding of the reactance of a typical coil, we need to review exactly what occurs when an a-c voltage is impressed across the coil.

1. The a-c voltage produces an alternating current.

2. When a current flows in a wire, lines of force are produced around the wire.

3. Large currents produce many lines of force; small currents produce only a few lines of force.

4. As the current changes, the number of lines of force will change. The field of force will seem to expand and contract as the current increases and decreases as shown in Figure 6.33.

5. As the field expands and contracts, the lines of force must cut across the wires that form the turns of the coil.

6. These cuttings induce an emf in the coil.

7. This emf acts in the direction so as to oppose the original voltage and is called a *counter*, or back, *emf.*

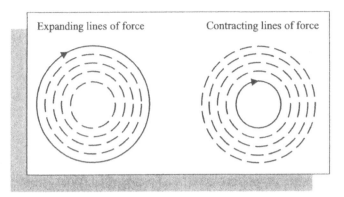

Expanding lines of force   Contracting lines of force

Figure 6.33
*An a-c current producing a moving (expanding and collapsing)
field. In a coil, this moving field cuts the wires of the coil.*

8. The effect of this counter emf is to reduce the original voltage impressed on the coil. The net effect will be to reduce the current below that which would flow if there were no cuttings or counter emf.

9. In this sense, the counter emf is acting as a resistance in reducing the current.

10. Although it would be more convenient to consider the current-reducing effect of a counter emf as a number of ohms of effective resistance, we don't do this. Instead, because a counter emf is not actually a resistance but merely **acts** as a resistance, we use the term *reactance* to describe this effect.

IMPORTANT

Important Point: *The reactance of a coil is the number of ohms of resistance, that the coil **seems** to offer as a result of a counter emf induced in it. Its symbol is X to differentiate it from the d-c resistance R.*

The inductive reactance of a coil depends primarily on (1) the coil's inductance and (2) the frequency of the current flowing through the coil. The value of the reactance of a coil is, therefore, proportional to its inductance and the frequency of the a-c circuit in which it is used.

The formula for inductive reactance is

$$X_L = 2\pi fL \qquad (6.23)$$

Because $2\pi = 2(3.14) = 6.28$, Equation (9.1) becomes

$$X_L = 6.28fL$$

where

X_L = inductive reactance, Ω
f = frequency, Hz
L = inductance, H

If any two quantities are known in Equation (6.23), the third can be found.

$$L = \frac{X_L}{6.28f} \qquad (6.24)$$

$$f = \frac{X_L}{6.28L} \qquad (6.25)$$

EXAMPLE 6.6

Problem: The frequency of a circuit is 60 Hz, and the inductance is 20 mh. What is X_L?

Solution:

$$X_L = 2\pi fL$$
$$= 6.28 \times 60 \times 0.02$$
$$= 7.5 \, \Omega$$

EXAMPLE 6.7

Problem: A 30-mh coil is in a circuit operating at a frequency of 1400 kHz. Find its inductive reactance.

Solution:

Given:
$$L = 30 \text{ mh}$$
$$f = 1400 \text{ kHz}$$
$$\text{Find } X_L = ?$$

Step 1: Change units of measurement.

$$30 \text{ mh} = 30 \times 10^{-3} \text{ h}$$
$$1400 \text{ kHz} = 1400 \times 10^3 \text{ Hz}$$

Step 2: Find the inductive reactance.

$$X_L = 6.28fL$$
$$X_L = 6.28 \times 1400 \times 10^3 \times 30 \times 10^{-3}$$
$$X_L = 263,760 \ \Omega$$

EXAMPLE 6.8

Problem:

Given:
$$L = 400 \ \mu\text{h}$$
$$f = 1500 \text{ Hz}$$
$$\text{Find } X_L = ?$$

Solution:

$$X_L = 2\pi fL$$
$$= 6.28 \times 1500 \times 0.0004$$
$$= 3.78 \ \Omega$$

Key Point: *If frequency or inductance varies, inductive reactance must also vary. A coil's inductance does not vary appreciably after* IMPORTANT *the coil is manufactured, unless it is designed as a variable inductor. Thus, frequency is generally the only variable factor affecting the inductive reactance of a coil. The coil's inductive reactance will vary directly with the applied frequency.*

6.7.2 CAPACITIVE REACTANCE

Previously, we learned that, as a capacitor is charged, electrons are drawn from one plate and deposited on the other. As more and more electrons accumulate on the second plate, they begin to act as an opposing voltage that attempts to stop the flow of electrons just as a resistor would do. This opposing effect is called the *reactance* of the capacitor and is measured in ohms.

The basic symbol for reactance is X, and the subscript defines the type of reactance. In the symbol for inductive reactance, X_L, the subscript L refers to inductance. Following the same pattern, the symbol for capacitive reactance is X_C are

 Key Point: *Capacitive reactance, X_C, is the opposition to the flow of a-c current due to the capacitance in the circuit.*

IMPORTANT

The factors affecting capacitive reactance, X_C, are:

- the size of the capacitor

- frequency

The larger the capacitor, the greater the number of electrons that may be accumulated on its plates. However, because the plate area is large, the electrons do not accumulate in one spot but spread out over the entire area of the plate and do not impede the flow of new electrons onto the plate. Therefore, a large capacitor offers a small reactance. If the capacitance were small, as in a capacitor with a small plate area, the electrons could not spread out and would attempt to stop the flow of electrons coming to the plate. Therefore, a small capacitor offers a large reactance. The reactance is therefore **inversely** proportional to the capacitance.

If an a-c voltage is impressed across the capacitor, electrons are accumulated first on one plate and then on the other. If the frequency of the changes in polarity is low, the time available to accumulate electrons will be large. This means that a large number of electrons will be able to accumulate, which will result in a large opposing effect, or a large reactance. If the frequency is high, the time available to accumulate electrons will be small. This means that there will be only a few electrons on the plates, which will result in only a small opposing effect, or a small reactance. The reactance is therefore **inversely** proportional to the frequency.

The formula for capacitive reactance is

$$X_C = \frac{1}{2\pi fC} \qquad (6.26)$$

with C measured in farads.

EXAMPLE 6.9

Problem: What is the capacitive reactance of a circuit operating at a frequency of 60 Hz, if the total capacitance is 130 μf?

Solution:

$$X_C = \frac{1}{2\pi fC}$$

$$= \frac{1}{6.28 \times 60 \times 0.00013}$$

$$= 20.4 \ \Omega$$

6.7.3 PHASE RELATIONSHIP R, L, AND C CIRCUITS

Unlike a purely resistive circuit (where current rises and falls with the voltage; that is, it neither leads nor lags and current and voltage are in phase), current and voltage are not in phase in inductive and capacitive circuits. This is the case, of course, because occurrences are not quite instantaneous in circuits that have either inductive or capacitive components.

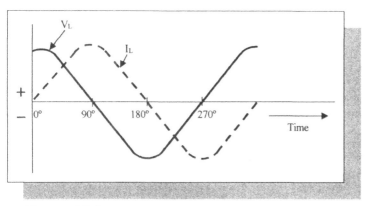

Figure 6.34
Inductive circuit—voltage leads current by 90°.

In the case of an inductor, voltage is first applied to the circuit, then the magnetic field begins to expand, and self-induction causes a counter current to flow in the circuit, opposing the original circuit current. Voltage **leads** current by 90° (see Figure 6.34).

When a circuit includes a capacitor, a charge current begins to flow, and then a difference in potential appears between the plates of the capacitor. Current **leads** voltage by 90° (see Figure 6.35).

Key Point: *In an inductive circuit, voltage leads current by 90°; and in a capacitive circuit, current leads voltage by 90°.*

IMPORTANT

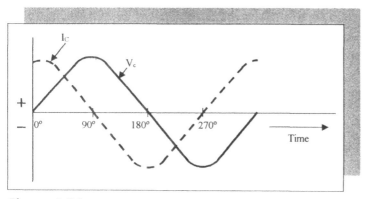

Figure 6.35
Capacitive circuit—current leads voltage by 90°.

6.7.4 IMPEDANCE

Impedance is the total opposition to the flow of alternating current in a circuit that contains resistance and reactance. In the case of pure inductance, inductive reactance, X_L is the total opposition to the flow of current through it. In the case of pure resistance, R represents the total opposition. The combined opposition of R and X_L in series or in parallel to current flow is called **impedance**. The symbol for impedance is Z.

The impedance of resistance in series with inductance is

$$Z = \sqrt{R^2 + X_L^2} \qquad (6.27)$$

where

Z = Impedance, Ω
R = Resistance, Ω
X_L = Inductive Reactance, Ω

The impedance of resistance in series with capacitance is

$$Z = \sqrt{R^2 + X_C^2} \qquad (6.28)$$

where

Z = Impedance, Ω
R = Resistance, Ω
X_C = Inductive Capacitance, Ω

When the impedance of a circuit includes R, X_L, and X_C, both resistance and net reactance must be taken into account. The equation for impedance, including both X_L and X_C, is

$$Z = \sqrt{R^2 + (X_L - X_C)^2} \qquad (6.29)$$

6.7.5 POWER IN REACTIVE CIRCUITS

The power in a d-c circuit is equal to the product of volts and amps, but, in an a-c circuit, this is true only when the load is resistive and has no reactance.

In a circuit possessing inductance only, the true power is zero. The current lags the applied voltage by 90°. The true power in a capacitive circuit is also zero. The *true power* is the average power actually consumed by the circuit, the average being taken over one complete cycle of alternating current. The *apparent power* is the product of the rms volts and rms amps.

The ratio of true power to apparent power in an a-c circuit is called the *power factor*. It may be expressed as a percent or as a decimal.

Note: *Earlier, we pointed out that the capacitive reactance decreases as the frequency increases and that inductive reactance*
IMPORTANT *increases as frequency increases. If a capacitor and an inductor are connected in series, there will be one frequency at which their reactances are equal. The frequency is called resonant frequency, which is covered in greater detail in Chapter 7.*

6.8 BASIC A-C CIRCUIT THEORY

Earlier, we explained how a combination of inductance and resistance and then capacitance and resistance behave in an a-c circuit. We saw how the *RL* and *RC* combination affect the current, voltages, power, and power factor of a circuit. We considered these fundamental properties as isolated phenomena. The following phase relationships were seen to be true:

1. The voltage drop across a resistor is **in phase** with the current through it.

2. The voltage drop across an inductor **leads** the current through it by 90°.

3. The voltage drop across a capacitor **lags** the current through it by 90°.

4. The voltage drops across inductors and capacitors are **180° out of phase**.

Solving a-c problems is complicated by the fact that current varies with time as the a-c output of an alternator goes through a complete cycle. This is the case because the various voltage drops in the circuit vary in-phase—they are not at their maximum or minimum values at the same time.

The a-c circuits frequently include all three circuit elements: resistance, inductance, and capacitance. In this section, all three of these fundamental circuit parameters are combined, and their effect on circuit values is studied.

6.8.1 SERIES RLC CIRCUITS

Figure 6.36 shows both the sine waveforms and the vectors for purely resistive (a), inductive (b), and capacitive (c) circuits. Only the vectors show the direction, because the magnitudes are dependent on the values chosen for a given circuit. [**Note:** We are only interested in the **effective** (root mean square, rms) values]. If the individual resistances and reactances are known, Ohm's Law may be applied to find the voltage drops. For example, we know that $E_R = I \times R$ and $E_C = I \times X_L$. Then, according to Ohm's Law, $E_L = I \times X_L$.

In an a-c circuit, current varies with time; accordingly, the voltage drops across the various elements also vary with time. However, the same variation is not always present in each **at the same time** (except in purely resistive circuits) because current and voltage are not in phase.

 Important Point: *In a resistive circuit, the phase difference between voltage and current is zero.*

IMPORTANT

We are concerned, in practical terms, mostly with effective values of current and voltage. However, to understand basic a-c theory, we need to know what occurs from instant to instant.

In Figure 6.37, note first that current is the common reference for all three element voltages, because there is only one current in a series circuit, and it is common to all elements. The dashed line in Figure 6.37(a) represents the common series current. The voltage vector for each element, showing its individual relation to the common current, is drawn above each respective element. The total source voltage E is the vector sum of the individual voltages of IR, IX_L, and IX_C.

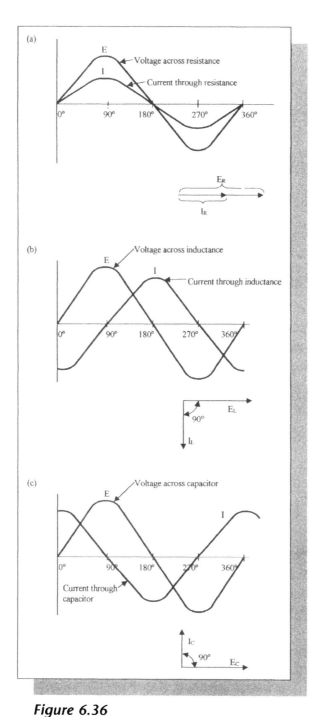

Figure 6.36
Sine waves and vectors for: (a) a pure resistive
circuit (voltage and current are in phase); (b) a pure
inductive circuit (voltage leads current by 90°); and (c)
a pure capacitive circuit (voltage lags current by 90°).

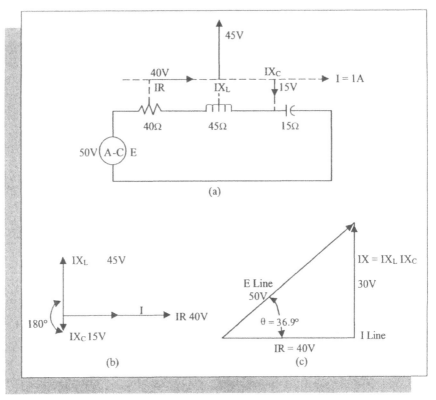

Figure 6.37
Resistance, inductance, and capacitance connected in a series.

The three element voltages are arranged for summation in Figure 6.37(b). Since IX_L and IX_C are each 90° away from I, they are 180° from each other. Vectors in direct opposition (180° out of phase) may be subtracted directly. The total reactive voltage E_X is the difference of IX_L and IX_C. Or, $E_X = IX_L - IX_C = 45 - 15 = 30$ volts.

 Important Point: *The voltage across a single reactive element in a series circuit can have a greater effective value than that of the applied voltage.*

IMPORTANT

The final relationship of line voltage and current, as seen from the source, is shown in Figure 6.37(c). Had X_C been larger than X_L, the voltage would lag, rather than lead. When X_C and X_L are of equal value, line voltage and current will be in phase.

Important Point: *One of the most important characteristics of an RLC circuit is that it can be made to respond most effectively*
IMPORTANT *to a single given frequency. When operated in this condition, the circuit is said to be in resonance with or resonant to the operating frequency. A circuit is at resonance when the inductive reactance X_L is equal to the capacitive reactance X_C. At resonance Z equals the resistance R.*

In summary, the series RLC circuit illustrates three important points:

1. The current in a series RLC circuit either leads or lags the applied voltage, depending on whether X_C is greater or less than X_L.

2. A capacitive voltage drop in a series circuit always subtracts directly from an inductive voltage drop.

3. The voltage across a single reactive element in a series circuit can have a greater effective value than that of the applied voltage.

6.8.2 PARALLEL RLC CIRCUITS

The **true power** of a circuit is $P = EI \cos \theta$; and for any given amount of power to be transmitted, the current, I, varies inversely with the power factor, $\cos \theta$. Thus, the addition of capacitance in parallel with inductance will, under the proper conditions, improve the power factor (make nearer to unity power factor) of the circuit and make possible the transmission of electric power with reduced line loss and improved voltage regulation.

Figure 6.38(a) shows a three-branch parallel a-c circuit with a resistance in one branch, inductance in the second branch, and capacitance in the third branch. The voltage is the same across each parallel branch, so $V_T = V_R = V_L = V_C$. The applied voltage V_T is used as the reference line to measure phase angle θ. The total current I_T is the vector sum of I_R, I_L, and I_C. The current in the resistance I_R is in phase with the applied voltage V_T (see Figure 6.38(b). The current in the capacitor I_C leads the voltage V_T by 90°. I_L and I_C are exactly 180° out of phase and, thus, are acting in opposite directions [see Figure 6.38(b). When $I_C > I_C$, I_T lags V_T [see Figure 6.38(c)] so the parallel RLC circuit is considered inductive.

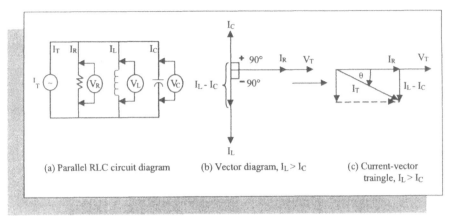

(a) Parallel RLC circuit diagram   (b) Vector diagram, $I_L > I_C$ (c) Current-vector traingle, $I_L > I_C$

Figure 6.38
R, X_L, and X_C in parallel.

6.9 POWER IN A-C CIRCUITS

In circuits that have only resistance, but no reactance, the amount of power absorbed in the circuit is easily calculated by $P = I^2R$. However, in dealing with circuits that include inductance and capacitance (or both), which is often the case in a-c electricity, the calculation of power is a more complicated process.

In Chapter 3, we explained that power is a measure of the rate at which work is done. The work of a resistor is to limit current flow to the correct, safe level. In accomplishing this, the resistor dissipates heat, and we say that power is consumed or absorbed by the resistor.

Inductors and capacitors also oppose current flow, but they do so by producing current that opposes the line current. In either inductive or capacitive circuits, instantaneous values of power may be very large, but the power actually absorbed is essentially zero, because only resistance dissipates heat (absorbs power). Both inductance and capacitance return the power to the source.

Any component that has resistance, such as a resistor or the wiring of an inductor, consumes power. Such power is not returned to the source, because it is dissipated as heat. Previously, we stated that power consumed in the circuit is called **true power**, or **average power**. The two terms are interchangeable, but we use the term *average power*, because the overall value is more meaningful than the instantaneous values of power appearing in the circuit during a complete cycle.

Key Point: *In terms of the dissipation of power as heat in a circuit,* **apparent power** *includes both power that is returned to the source and power that is dissipated as heat.* **Average power** *is power that is dissipated as heat.*

IMPORTANT

Not all apparent power is consumed by the circuit; however, because the alternator does deliver the power, it must be considered in the design. The average power consumption may be small, but instantaneous values of voltage and current are often very large. Apparent power is an important design consideration, especially in assessing the amount of insulation necessary.

In an a-c circuit that includes both reactance and resistance, the load consumes some power, and some is returned to the source. How much of each depends on the phase angle, because current normally leads or lags voltage by some angle.

Note: *Recall that in a purely reactive circuit, current and voltage are 90° out of phase.*

IMPORTANT

In an RLC circuit, the ratio of R/Z is the cosine of the phase angle θ. Therefore, it is easy to calculate average power in an RLC circuit:

$$P = EI \cos \theta \qquad (6.30)$$

where

E = effective value of the voltage across the circuit
I = effective value of current in the circuit
θ = phase angle between voltage and current
P = average power absorbed by the circuit

Note: *Recall that the equation for average power in a purely resistive circuit is P = EI. In a resistive circuit, P = EI, because the* $\cos \theta$ *is 1 and need not be considered. In most cases, the phase angle will be neither 90° nor 0°, but somewhere between those extremes.*

IMPORTANT

EXAMPLE 6.10

Problem: An RLC circuit has a source voltage of 500 volts, line current is 2 amps, and current leads voltage by 60°. What is the average power?

Solution:

$$Average\ Power = 500\ V \times 2\ A \times 0.5$$
$$(Note:\ cos\ of\ 60° = 0.5)$$
$$= 500\ watts$$

EXAMPLE 6.11

Problem: An RLC circuit has a source voltage of 300 volts, line current is 2 amps, and current lags voltage by 31.8°. What is the average power?

Solution:

$$Average\ Power = 300\ V \times 2\ A \times 0.8499$$
$$= 509.9\ watts$$

EXAMPLE 6.12

Problem:

Given:
$$E = 100\ volts$$
$$I = 4\ amps$$
$$\theta = 58.4$$

What is the average power?

Solution:

$$Average\ Power = 100\ V \times 4\ A \times 0.5240$$
$$= 209.6\ watts$$

Self-Test

6.1 As the armature of an alternator moves through one complete rotation, what must occur for maximum voltage to be generated?

6.2 What is electromagnetic induction?

6.3 What is a sine wave?

6.4 What two values does a vector represent?

6.5 The rms voltage generated by an alternator is 600 volts. What is the peak voltage?

6.6 When an a-c voltage is impressed across a coil, the resulting current is an _____ current. This changing current produces changing fields of force that _____ the wires of the coil. These cuttings induce a _____ emf in the coil.

6.7 What characteristics of electric current result in self-inductance?

6.8 List four factors affecting the inductance of a coil.

6.9 An increase in the length of a coil will _____ inductance.

6.10 Name the three factors that affect capacitance.

6.11 Define the *RC* time constant.

6.12 A capacitor is used to store up an electric _____.

6.13 The reactance of a capacitor is the opposition offered by the capacitor to the passage of an _____ current.

6.14 What are the factors affecting inductive reactance?

6.15 What is the phase relationship between voltage and current in a capacitive circuit?

6.16 Define true power.

6.17 What kind of power is associated with purely resistive circuits?

6.18 In a purely reactive circuit, current and voltage are _____ degrees out of phase.

6.19 What is the cos θ in a purely resistive circuit?

6.20 Define power factor.

Resonance

TOPICS

The Resonance Effect
Series Resonance
Parallel Resonance
Q Factor
Bandwidth
Tuning

Key Terms Used in This Chapter

BANDWIDTH	A range of frequencies that have a resonant effect.
PARALLEL RESONANT CIRCUIT	A circuit including a capacitor, an inductor, and sometimes a resistor, connected in parallel, and in which the inductive and capacitive reactances are equal at the applied-signal frequency. The circuit impedance is a maximum, and the current through the circuit is a minimum at the resonant frequency.
Q OF RESONANT CIRCUIT	Figure of quality or merit, in terms of reactance compared with resistance.
SERIES RESONANT CIRCUIT	A circuit including a capacitor, an inductor, and sometimes a resistor, connected in series, and in which the inductive and capacitive reactances are equal at the applied-signal frequency. The circuit impedance is at a minimum, and the current is a maximum at the resonant frequency.
TUNING	Varying the resonant frequency of an LC circuit.

7.1 INTRODUCTION

In Chapter 6, we saw how the inductor and capacitor each present an opposition (reactance) to the flow of an a-c current and how the magnitude of this reactance depends upon the frequency of the applied signal.

When combined, these two components, in series or parallel, have a definite and profound effect, and a peculiar phenomenon results: *resonance*. The resonance effect occurs when the inductive and capacitive reactances are equal.

The primary application of resonance is in radio frequency (RF) circuits for tuning to an a-c signal of the desired frequency (e.g., as used in tuning in radio and television receivers, transmitters, and electronics equipment).

In this chapter, we look at some of the properties of resonant circuits and concentrate on those properties important in the operation of oscillators (covered in Chapter 9).

7.2 THE RESONANCE EFFECT

What is resonance effect? Probably the best way to answer this question is to begin with a few basic but important facts with which we should be familiar with.

1. Inductive and capacitive reactances (X_L and X_C) have *opposite* effects on a circuit.

2. The voltage across an inductor is 90° out of phase with the current through the inductor. The current through a capacitor is 90° out of place with the voltage across it.

3. The voltage leads the current in an inductor and lags the current in a capacitor.

4. In a series circuit, the inductor voltage and capacitor voltage are 180° out of phase.

5. In a parallel circuit, the inductor current and capacitor current are 180° out of phase.

6. When we decrease signal frequency, inductive reactance decreases. On the other hand, when we decrease frequency, capacitive reactance increases, and opposite effects occur. If we increase the frequency, opposite effects also occur to inductive and capacitive reactance. Specifically, inductive reactance increases and capacitive reactance decreases when we increase the frequency.

The preceding statements summarize the basic relationship between increases or decreases in circuit frequency: the direct and inverse impact upon inductive and capacitive reactance.

We still need to answer the question: What is resonance effect? Consider Figure 7.1—a graph of inductive reactance for a coil and the capacitive reactance for a capacitor. The line for inductive reactance increases as we change the frequency. The line for capacitive reactance indicates that, as we increase the frequency, the capacitive reactance decreases. Note that we did not include a frequency or a reactance scale for Figure 7.1. The fact is, scales are not necessary—exact values are unimportant. The important point illustrated in Figure 7.1 is that for any capacitor and inductor, their reactance lines will cross at only one point.

Figure 7.1 illustrates the special condition that occurs when inductive and capacitive reactances are equal. This is true whether we have a series or a parallel circuit. We call this condition *resonance effect*.

Any LC circuit can exhibit resonance effect. It all depends on the frequency; that is, frequency is the key parameter. At the resonant frequency,

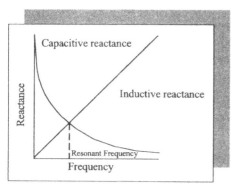

Figure 7.1
Graph of inductance reactance of a coil and of capacitive reactance. For any capacitor and inductor, their reactance lines cross at only one point—when reactances are equal. We call this condition resonance.

an LC combination provides the resonance effect. When either below or above resonant frequency, the LC combination is just another a-c circuit.

 Key Point: *Every combination of inductance and capacitance has one frequency at which the reactances are equal. This is the* **resonant frequency** *for the combination.*

IMPORTANT

The frequency at which the opposite reactances are equal—the resonant frequency—can be calculated as:

$$f_r = \frac{1}{2\pi\sqrt{LC}} \qquad (7.1)$$

where

L = inductance in henrys
C = capacitance in farads
f_r = resonant frequency in hertz

Because $1/2\pi$ is a numerical value equal to $1/6.28$, or 0.159, a more convenient calculation is

$$f_r = \frac{0.159}{\sqrt{LC}} \qquad (7.2)$$

We can use this equation to calculate the resonant frequency of any combination of inductor and capacitor values. We also can solve Equations (7.1) and (7.2) for values of inductance and capacitance. Given either an inductor or a capacitor, we can calculate the other component value to give a desired resonant frequency.

EXAMPLE 7.1

Problem: Calculate the resonant frequency for a 20-Hz inductor and an 8-μF capacitor.

Solution:

$$f_r = \frac{1}{2\pi\sqrt{LC}}$$

$$= \frac{0.159}{\sqrt{20 \times 8 \times 10^{-6}}}$$

$$= \frac{0.159 \times 10^{3}}{\sqrt{160}}$$

$$= \frac{159}{12.65}$$

$$f_r = 12.6 \text{ Hz}$$

Key Point: *For any LC circuit (series or parallel), $f_r = 1/(2\pi \sqrt{LC})$ is the resonant frequency that makes the inductive and capacitive* IMPORTANT *reactances equal.*

7.3 SERIES RESONANCE

When the reactance in a series circuit cancels, and, as far as the source is concerned, the circuit appears to contain only resistance, the circuit is in a condition of **resonance**. When resonance is established in a series circuit, certain conditions prevail:

1. The current I is maximum at the resonant frequency f_r.

2. The current I is in phase with the generated voltage, or the phase angle of the circuit is $0°$.

3. The voltage is maximum across either L or C alone.

4. The impedance is minimum at f_r, equal only to the low series resistance.

Figure 7.2 shows a simple series resonant circuit.

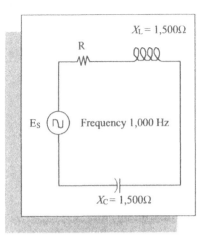

Figure 7.2
A single series resonant circuit.

7.4 PARALLEL RESONANCE

In the pure, or ideal, parallel resonant circuit (i.e., one in which there is no resistance), the inductor and capacitor are placed in parallel, and the applied voltage E_T appears across these circuit components (see Figure 7.3). In this parallel resonant circuit, as in the series resonant circuit, resonance occurs when the inductive reactance is equal to the capacitive reactance

The main characteristics for a parallel resonant circuit are

1. The line current I_T is minimum at the resonant frequency.

2. The current I_T is in phase with the generator voltage E_A, or the phase angle of the circuit is $0°$.

3. The impedance Z_T, equal to E_A/I_T, is maximum at f_r because of the minimum I_T.

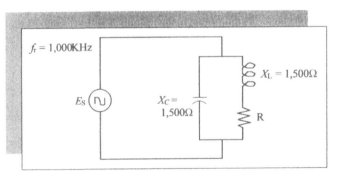

Figure 7.3
Parallel resonant circuit.

7.5 Q FACTOR

The quality, or magnification factor, or figure of merit, of the resonant circuit, in sharpness of resonance, is indicated the factor Q. This Q factor is a number assigned to a coil or capacitor to indicate the component's relative merits. An entire circuit also can have a Q value. A circuit Q value gives an idea of how close to the ideal that circuit performs.

Q is defined as the ratio of reactance to resistance. This definition applies to both inductors and capacitors.

The Q of an ideal inductor, L, is equal to the inductive reactance divided by the resistance. Equation (7.3) is the mathematical statement of this definition.

$$Q = \frac{X_L}{R} \qquad (7.3)$$

The Q of an ideal capacitor, C, is equal to the capacitive reactance divided by the resistance. Equation (7.4) states this definition mathematically.

$$Q = \frac{X_C}{R} \qquad (7.4)$$

 Important Point: *The obvious thing about these equations is that smaller resistance values give larger Q values. Larger Q values* IMPORTANT *indicate better-quality components.*

7.6 BANDWIDTH

We pointed out that an LC circuit is resonant at "one" frequency— this is the case for the maximum resonance effect. However, other frequencies close to resonant frequency, or f_r, also are effective. The fact is, it is practically impossible to have an LC circuit with a resonant effect at only one frequency. For example, for series resonance, frequencies just below or above f_r produce increased current, but a little less than the value of resonance. Similarly, for parallel resonance, frequencies close to f_r can provide a high impedance, although a little less than the maximum Z_T.

This means that, at any resonant frequency, there is an associated band of frequencies present that provide resonance effects. The actual width of the band depends on the Q of the resonant circuit.

The width of the resonant band of frequencies centered around f_r is called the *bandwidth* of the resonant (or tuned) circuit.

Key Point: *The width of the resonant band of frequencies centered around f_r is called the **bandwidth** of the circuit.*

IMPORTANT

Figure 7.4 illustrates exactly what we mean by bandwidth. In the figure, the group of frequencies with a response of 70.7% of maximum or more is considered the bandwidth of the tuned circuit. For a series resonant circuit, the bandwidth is measured between the two frequencies f_1 and f_2, producing 70.7% of the maximum **current** at f_r. For a parallel resonant circuit, the bandwidth is measured between the two frequencies, allowing 70.7% of the **maximum total impedance** at f_r.

Key Point: *In a series resonant circuit, the resonant response increases current. For a parallel resonant circuit, the resonant*
IMPORTANT *response is an increased impedance.*

The group of frequencies with a response 70.7% of maximum, or more, is generally considered the bandwidth of the turned circuit, as shown in Figure 7.4.

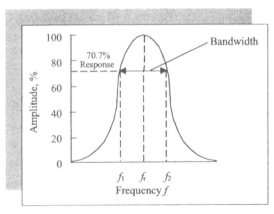

Figure 7.4
Bandwidth of a tuned LC circuit.

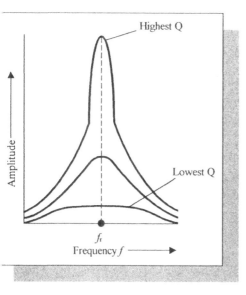

Figure 7.5
Resonant response curve.

Bandwidth in terms of Q is

$$BW = \frac{f_r}{Q}$$

where

$$Q = \frac{X_L}{R}$$

The Q of a circuit is a measure of its bandwidth. High Q means narrow bandwidth, whereas low Q yields broader resonant response and greater bandwidth (see Figure 7.5).

Important Point: *Generally, a high Q is desirable for more output from the resonant circuit. However, it must be enough bandwidth to include the desired range of signal frequencies.*

7.7 TUNING

The most common application of resonance in radio frequency (RF) circuits is called *tuning*. In this use, the LC circuit provides maximum voltage output at the resonant frequency, composed with the amount of output at any other frequency either below or above resonance.

Generally, we can say that any circuit operated to provide frequency selectivity is a **tuned circuit**.

Note: *All examples of tuning in radio and television are applications of resonance. Generally, tuned circuits are used in impedance matching, bandpass filters, and oscillators.*

IMPORTANT

Important Point: *When an LC circuit is tuned, the change in resonant frequency is inversely proportional to the square root of the change in LC. For electronic tuning, the capacitance (C) is varied by a* ***varactor.***

IMPORTANT

Self-Test

7.1   What is the voltage across the resistor in a series RLC circuit at the resonant frequency?

7.2   What is the voltage across a resistor in series with a parallel LC circuit at the resonant frequency?

7.3   What is the impedance of a series circuit at resonance?

7.4   Series resonance is the condition at which the inductive reactance is exactly equal to the capacitive reactance. For any given combination of coil and capacitor, there is only one frequency at which this situation can occur. This frequency is called the _____ frequency.

7.5   For series resonance, impedances for X_L and X_C are _____ and _____ .

7.6   For series resonance, current is _____ .

7.7   For parallel resonance, current is _____ .

7.8   At resonance, the phase angle equals _____ .

7.9   Define bandwidth.

7.10   At resonance, the total impedance is considered to be all _____ .

Transistor Amplifiers

TOPICS

Amplifier Basics
Transistor Amplifiers
Single-Stage Amplifiers
Multistage Amplifiers
Op Amps

8.1 INTRODUCTION

To amplify means to increase: to make larger or more powerful. An amplifier is a circuit that increases (amplifies) signal strength. Most a-c signal inputs to amplifiers are very small (feeble) and provide a signal from low-level devices (for example, the output from a record pick-up or from microphones in public address systems). Amplifiers increase the signal strength to the power level needed to operate these devices so their signals can be heard. Amplifiers are used in many kinds of electronic circuits, including computers, television, digital watches, microwave ovens, and pocket calculators.

In this chapter, we emphasize the basic principles of amplifier operation, how op amps work, and how they are used. We also discuss amplifier characteristics, signal distortion, and common kinds of amplifiers.

We present the simplest and most basic of the possible amplifying circuits, to demonstrate how transistors amplify a signal and to provide a basic foundation in transistor amplifiers for more advanced studies.

Key Terms Used in This Chapter

AMPLIFIER	A device that increases the amplitude of a signal.
GAIN	Also amplification. Ratio of amplified output to input.
HALF-POWER POINTS	Those points on the response curve of a resonant circuit where the power is one half its value at resonance.
EMITTER-FOLLOWER	Circuit in which signal input is to base and output is from emitter. Same as common-collector circuit.
BIASING CIRCUIT	A sub-circuit that establishes the operating point of an amplifier circuit.
SATURATION POINT	Maximum limit at which changes of input have no control in changing the output.
CUTOFF POINT	The point at which base current equals zero and the collector-emitter voltage is at its maximum possible.
POWER GAIN	The ratio of the signal power delivered to its load to the signal power adsorbed by its input circuit.
EFFICIENCY	Ratio of power output to power input.
CURRENT GAIN	The ratio of current output to current input.
VOLTAGE GAIN	The ratio of voltage output to voltage input.
COUPLING	Joining two or more circuits to pass a signal from one to another.
OPERATIONAL AMPLIFIER (OP AMP)	High-gain amplifier with negative feedback, commonly used in linear IC chips and A/D convertors.
DIFFERENTIAL AMPLIFIER	A device that amplifies the voltage difference between two inputs.

8.2 AMPLIFIER BASICS

Producing an output greater than its input is what an *amplifier* is all about. Voltage, current, or both may be amplified. Because power is the product of current and voltage, it, too, is amplified. Transistors, vacuum tubes, and op amps, if fed from a proper power supply, are examples of amplifiers.

Those familiar with transformers (specifically, transformer action) may wonder why we did not include the transformer as an amplifier. While true that a step-up transformer, for example, can increase voltage, it is also true that it cannot, however, increase power, because current decreases proportionately to the increase in voltage. The bottom line: the transformer does not amplify or increase power. We cover transformers in greater detail in Chapter 11.

8.2.1 SIMPLE TRANSISTOR AMPLIFIER OPERATION

Figure 8.1 shows a simple NPN transistor amplifier. The power source is a d-c voltage (+E at the top of the figure). The input signal is applied to the base of the transistor, and the output is taken from the collector.

When a small voltage is applied to the amplifier shown in Figure 8.1, a small voltage is produced across the base-emitter junction. This small

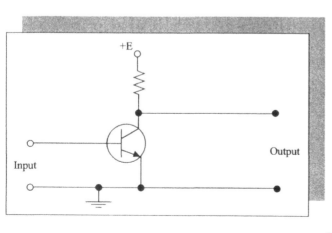

Figure 8.1
Simple NPN
transistor amplifier.

voltage in turn causes a small base current, which follows the input and controls the collector current. The output collector current is much greater than the base current. This is an important point because power is equal to the product of voltage and current, and thus the amplifier output is greater than its input power.

IMPORTANT

> **Key Point:** *From the simple NPN transistor circuit shown in Figure 8.1 and description provided above, it should be apparent that the purpose of an amplifier circuit is to produce an output greater than its input.*

8.2.2 AMPLIFIER CHARACTERISTICS

Amplifier circuits are typically designed with certain characteristics, making them suitable for specific applications but not for others. The three most important of these characteristics are gain, bandwidth, and distortion.

8.2.2.1 GAIN

We have shown that an amplifier increases the amplitude of an electronic signal. The ratio of output to input is referred to as *gain*. Gain can be expressed in three ways: as **voltage gain**, as **current gain**, or as **power gain** (the product of voltage and current). Mathematically this is expressed as:

$$\textbf{gain} = \frac{\textbf{output}}{\textbf{input}} \qquad (8.1)$$

For example, if an amplifier has an output signal of 4 and an input signal of 1, the gain is 4, or four times the input.

In certain applications, the use of a single amplifier will not provide enough gain to amplify a signal to the desired level. In this case, several amplifiers (or stages) can be used in a *cascade* arrangement, in which the output of one amplifier is used as the input to another. As shown in Figure 8.2, the amplifiers Q1, Q2, Q3, and Q4 are in cascade.

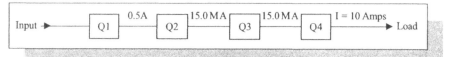

Figure 8.2
Amplifier stages in cascade. Note the increase in amperage for each successive transistor stage.

IMPORTANT

Key Point: Amplifiers connected with the output driving the input of the next stage are cascaded stages.

8.2.2.2 BANDWIDTH

It is important to note that the gain of an amplifier is not the same for all signal frequencies. An amplifier is designed to operate within a certain range of frequencies, referred to as *bandwidth*. (Note: Bandwidth was covered earlier in Section 7.6.) Specifically, an amplifier operates between a lower critical frequency and an upper critical frequency, and the difference between the two critical frequencies is its bandwidth. Maximum voltage gain occurs at the **midband** (the middle range of frequencies) and decreases both above and below the midband.

Earlier (in Section 7.6), we described the bandwidth as being that region between two frequencies having 70.7% response. This is simply for convenience in calculation. Recall that at each of the two frequencies, the net capacitive and inductive reactance equals the resistance. This means that the total impedance of the series reactance and resistance is 1.4 times greater than the resistance. With this much more Z_T, the current is reduced to 1/1.414, or 0.707, of its maximum value.

Moreover, the relative current or voltage value of 70.7% corresponds to 50% in power, because power is E^2/R or I^2/R and the square of 0.707 equals 0.50 [as shown in Equation (8.2)]. Thus, the bandwidth between frequencies having 70.7% response in current or voltage is also the bandwidth in terms of *half-power points*. That is, the lower and upper frequency limits are defined by the power points.

$$\text{Power Output} = 0.707\, E_{\text{Max}} \times 0.707\, I_{\text{Min}} = 0.50\, P_{\text{Max}} \quad (8.2)$$

where

E_{Max} = Maximum Output Voltage
I_{Min} = Minimum Output Current
P_{Max} = Maximum Output Power

8.2.2.3 DISTORTION

Unwanted changes in the output signal waveforms are referred to as *distortion*. Whether distortion occurs in the frequency (when frequencies in the input signal are amplified more than other frequencies), the amplitude (when amplitude changes in the input signal do not produce proportional amplitude changes in the output signal), or the phase of an output signal (when different frequencies in the input signal are shifted in phase by different amounts in the output), it does occur. Consequently, amplifiers do not reproduce the input signal exactly.

Key Point: *In the selection of an amplifier for a particular application, distortion is an important factor that must be considered.*

IMPORTANT

8.3 TRANSISTOR AMPLIFIERS

Earlier, in Chapter 5, we described transistors and their operation. The two kinds of transistors we described are used in amplifiers: (1) *bipolar transistors* (current-controlled devices) and (2) *field-effect transistors* (FETs—voltage-controlled devices). The amplifier circuits are very similar—identical, in some cases—for the two kinds of transistors.

Let's review transistor basics—the differences and similarities between bipolar transistors and FET's—before we launch into a more detailed discussion of transistor amplifiers.

Recall that current in the bipolar transistor depends on the flow of both majority and minority carriers, but current in an FET depends on majority carriers only. The three terminals of the bipolar transistor are the base, emitter, and collector. In the bipolar NPN transistor, the emitter-base junction is forward-biased, and the collector-base junction is reverse-biased. In FETs, a channel between the source and the drain carries a current in response to a small voltage at the gate.

8.3.1 BIPOLAR TRANSISTOR AMPLIFIERS

The current through the transistor, the voltage across the transistor, and the transistor power gain are all dependent on the manner in which the transistor is connected in a circuit. With the exception of reversed polarities, the input and output connections from NPN transistors are the same as PNP transistors. The three basic connections—common emitter, common base, and common collector (emitter follower)—are shown in Figure 8.3.

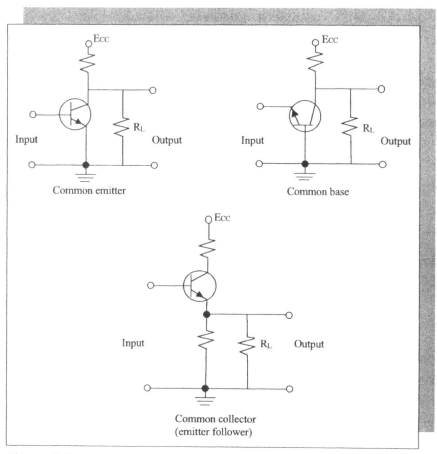

Figure 8.3
Bipolar transistor connections.

TABLE 8.1. Comparison of Transistor Circuits.			
Characteristics	Common Base	Common Emitter	Common Hollector
Current gain	Less than 1	High	High
Voltage gain	High	High	Less than 1
Power gain	Medium	Very high	Medium
Phase inversion	No	Yes	No

IMPORTANT **Important Point:** *The kind of transistor circuit connection is named for the transistor element at a-c ground (e.g., the common-emitter circuit has input to the base, output from the collector, and the emitter to ground).*

Important characteristics of each bipolar transistor amplifier connection are listed in Table 8.1.

The characteristics of the bipolar transistor amplifier connections listed in Table 8.1 clearly indicate that the function of the amplifier in the application determines how it should be connected. The common-emitter connection is a case in point. For example, this connection provides the highest power gain and has the characteristic of inverting the signal, so we would use this connection when these features are important to the application.

8.3.2 FIELD-EFFECT TRANSISTOR (FET) AMPLIFIERS

The FET's main advantage over bipolar transistors is its very high output impedance. Figure 8.4 indicates that FET amplifier connections are very similar to bipolar transistor amplifier connections; however, the terminology used is different. The three FET elements are called the **source** (corresponds to the emitter), the **gate** (corresponds to the base), and the **drain** (corresponds to the collector). As shown in Figure 8.4, basic FET amplifier connections are referred to as common source, common gate, and common drain.

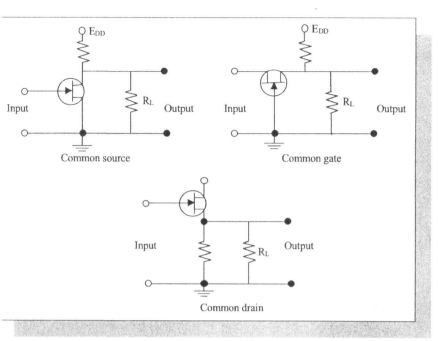

Figure 8.4
FET amplifier connections.

Important Point: A bipolar transistor is controlled by current, and an FET is controlled by voltage.

IPORTANT

8.4 SINGLE-STAGE AMPLIFIERS

[Note: For discussion purposes, and to facilitate ease of explanation of single-stage amplifiers, we have selected two simple, single-stage audio amplifiers (one using a bipolar transistor and the other using an FET transistor) for our discussion in the Sections 8.4.1 and 8.4.2.]

Audio amplifier circuits are designed to amplify signals that lie within the audio frequency spectrum (20 to 20,000 Hz). In practice, to ensure that these signals are amplified to sufficient power levels, audio amplifiers are designed to respond to frequencies above and below the audio frequency. Transistors are used in audio frequency amplifiers for voltage or power amplification.

8.4.1 SINGLE-STAGE AUDIO AMPLIFIER (BIPOLAR TRANSISTOR)

Figure 8.5 shows a basic, single-stage audio amplifier using a transistor as the amplifying device. The input will be a low amplitude audio voltage (millivolt range) that will be amplified by Q1 and coupled either to another voltage amplifier or, if sufficient amplification has been obtained in this stage, to a power amplifier. The circuit shown in Figure 8.5 uses a PNP transistor in the common-emitter connection. R_L serves as the collector load resistor, R_E and C_E are the emitter bias stabilization network, and R_B limits the bias current and thereby establishes the operating point. Coupling capacitor C2's value is such so as to ensure good frequency response. When $-E_{CC}$ is applied to the circuit, the current flow through the base-emitter junction develops a voltage that forward biases the input circuit.

During the positive alternation of the input signal, forward bias is being decreased. This causes a resulting decrease in collector current, and the collector voltage becomes more negative (with less of a voltage drop across R_L, the collector voltage will increase toward the negative value of E_{CC}). The output signal at the collector is an amplified, negative-going signal.

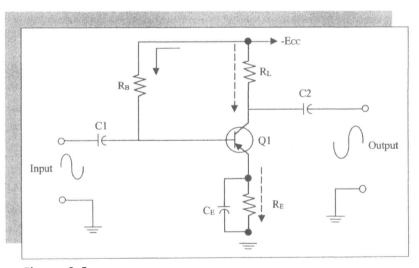

Figure 8.5
PNP transistor audio amplifier.

During the negative alternation of the input signal, the forward bias is increased. This increases the collector current through R_L, with the result that the collector voltage becomes less negative (more positive). The output signal during this half cycle is an amplified positive half cycle. The paths for base and collector currents (d-c) are shown in Figure 8.5. The dotted path is for collector current, and the solid line shows the base biasing current.

Important Point: *In order to attain a maximum transfer of energy, the input impedance of the transistor should be as close as* IMPORTANT *possible to the internal impedance of the device being used as the voltage source. Simply stated, the maximum transfer of energy from one circuit to another will occur when the impedances of the two circuits are equal or* **matched.** *This standard procedure is called impedance matching.*

8.4.2 SINGLE-STAGE AUDIO AMPLIFIER (FET)

Figure 8.6 shows a basic FET (field effect transistor) audio amplifier. This circuit utilizes an N-channel FET as the amplifying device. Recall that the FET is a high gain device. The circuit shown utilizes the

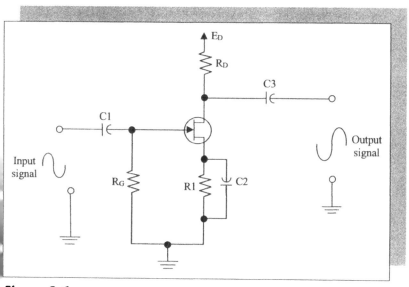

Figure 8.6
FET audio amplifier.

common-source connection. Self-bias is developed by the source-bias circuit consisting of R1 and C2. C1 is the input coupling capacitor, R_G is the input gate resistance used to develop the input signal, C3 is the output coupling capacitor, and R_D is the drain load resistor. In Figure 8.6, as the gate becomes less negative with respect to the source, source-drain current will increase. As the gate becomes more negative with respect to the source, source-drain current will decrease.

During the positive alternation of the input signal, the gate becomes less negative than its quiescent (at rest) voltage, the depletion region decreases, and drain current increases. The increase in drain current is proportional to the change in junction bias. As drain current increases, causing a corresponding decrease in drain voltage, output voltage is consequently decreasing, or passing through its negative half cycle. As shown in Figure 8.6, the common source connection provides a 180° phase inversion.

During the negative alternation of the input signal, the gate voltage becomes more negative than its quiescent value, and the depletion region increases causing drain current to decrease. The decreased drain current will drop less voltage across R_D resulting in an increased drain voltage. Thus, the output voltage is going through its positive alternation.

FETs have a high input impedance that immediately makes them acceptable for use with high impedance input devices. Their output impedances are rated as medium to high and are, therefore, applicable for driving high-impedance earphones or high power amplifiers. As a rule, FET circuits are less complex than standard transistor circuits as they require fewer circuit components.

8.5 MULTISTAGE AMPLIFIERS

Generally, a single-stage amplifier is used to provide a low level of amplification (e.g., for a small public address system). If more amplification is needed, two or more single-stage amplifiers can be connected together to form a *multistage amplifier*. Multistage amplifiers are also known as **cascade amplifiers**. They can be thought of as chains of single-stage amplifiers, as shown in Figure 8.7.

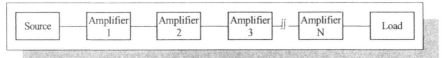

Figure 8.7
Chain of single-stage amplifiers forming a multistage amplifier.

 Key Point: Single-stage amplifiers are often joined to build a multistage amplifier. Multistage amplifiers provide greater amplification
IMPORTANT than a single unit can provide.

Figure 8.8 shows a basic multistage amplifier using two common-emitter amplifiers. E_S is the source for amplifier 1, and amplifier 2 is its load. Amplifier 1 is the source for amplifier 2. The gain of the multistage amplifier is the product of the gain of amplifier 1 and the gain of amplifier 2, or

$$\text{Total Gain} = \text{Amplifier 1 gain} \times \text{Amplifier 2 gain} \qquad (8.3)$$

For example, suppose the gain of amplifier 1 is 10, and the gain of amplifier 2 is 20. The total gain of the circuit shown in Figure 8.8 is then $10 \times 20 = 200$.

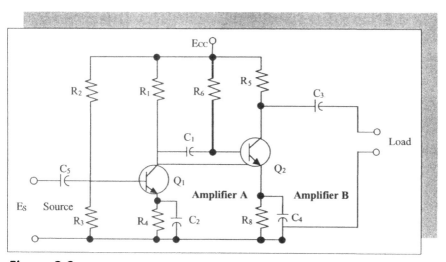

Figure 8.8
Basic multi-stage amplifier.

Key Point: *Total gain is the product of the individual amplifier gains for power, current, and voltage gain.*

IMPORTANT

Earlier, we pointed out that amplifier **bandwidth** is the range of frequencies over which gain is nearly constant. The half-power points determine the upper and lower frequencies of the bandwidth. Figure 8.9 shows the frequency-response curve for a typical amplifier. The half-power points (P_1 and P_2) are those points at which the power output is half the maximum power output. The shaded area shows the amplifier bandwidth.

In Figure 8.9, the frequency at P_1 is called the *lower half-power frequency*. It is the low-frequency limit of the bandwidth. The lower half-power frequency generally is greater for a multistage amplifier than it is for any one of its stages. Similarly, the *upper half-power frequency*, P_2, is usually less for a multistage amplifier than it is for any one of its stages. Therefore, the bandwidth of a multistage amplifier always is less than the bandwidth of any of its stages.

8.5.1 MULTISTAGE COUPLING

Coupling means connecting the output of one circuit to the input of the next. In amplifier coupling, the goal is to join, or couple, two or more circuits in order to pass a signal from one to another. The point is that usually more than one amplifier, operated in cascade, is needed to increase the amplitude of the feeble input signal to the required output value. Three

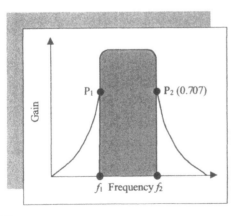

Figure 8.9
Frequency response curve.

basic methods of coupling amplifier stages to each other and to loads are

● capacitive coupling

● transformer coupling

● direct coupling

When coupling amplifier stages, it is important to note that no one coupling method is best for all applications, and the advantages and disadvantages of each must be weighed. Important considerations include frequency response, impedance matching, cost, and the physical size and weight of the circuit.

Important Point: *While all coupling networks are frequency responsive, some coupling methods provide better response than others, but their basic principles of operation remain the same regardless of whether or not they are employed singly as input or output coupling devices, or in cascade.*

8.5.1.1 CAPACITIVE COUPLING

The circuit shown previously in Figure 8.8 is a capacitive-coupled two-stage amplifier. (Note: This method is probably the most common type of coupling in amplifier circuits—this is the case even though capacitive coupling limits low-frequency response and does nothing to aid impedance matching.) In this circuit, the coupling between amplifier stages, the source, and the load is by capacitors. Capacitors C_1, C_2, and C_3 isolate each stage from the other by presenting an open circuit to the flow of d-c. The requirements are to include all frequencies in the desired signal, while rejecting undesired components. Usually, the d-c component must be blocked from the input to a-c amplifiers. The purpose is to maintain a specific d-c level for the amplifier operation—only an a-c signal is allowed to pass from source to load.

Important Point: *Capacitive coupling passes an a-c signal only.*

8.5.1.2 TRANSFORMER COUPLING

Interstage transformers can be used to match input and output impedances between stages. Transformers also are used to match load impedance
to amplifier output impedance. Figure 8.10 shows a transformer-coupled
two-stage amplifier.

The interstage, or coupling, transformer (T_1) couples the signal from
the collector circuit of Q_1 to the base of Q_2. A transformer, T_2, also couples the load.

Transformer coupling in general is very efficient. However, the size,
weight, and cost of transformers make them unacceptable in many applications. Also, the bandwidth, at higher frequencies, is narrower than in
capacitive-coupled circuits. The limitations of frequency response generally restrict the use of transformer coupling to audio circuits that do not
require an exceptionally wide bandpass or frequency response, but do require voltage or power outputs.

Key Point: *Transformer coupling is very efficient, but its high-
frequency response is limited.*

IMPORTANT

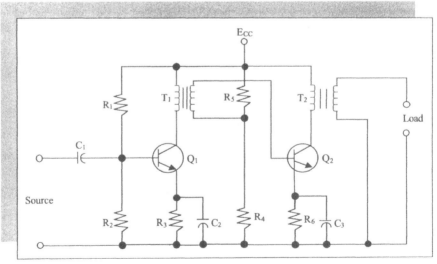

Figure 8.10
Transformer-coupled amplifier.

8.5.1.3 DIRECT COUPLING

Because both capacitive-coupled and transformer-coupled amplifiers block low-frequency signals, another method is used for applications with very low frequencies. The direct-coupled amplifier provides a response that extends down to 0 Hz (d-c).

Figure 8.11 shows a basic three-stage direct-coupled amplifier. The main function of this circuit is to amplify signals below several hertz. Direct-coupled circuitry is often used in computer circuits and in the output circuits of video amplifiers because of its ability to amplify direct current or zero frequency. The circuit is capable of extremely high current gain, but is quite sensitive to an undesired change in output with constant input; this phenomenon is called *drift*. The main causes of drift are variations in the power supply and changes in the transistor characteristics due to temperature or age.

Key Point: *Direct-coupled amplifiers have good low-frequency response, but they are sensitive to drift.*

IMPORTANT

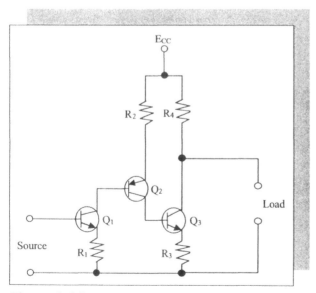

Figure 8.11
Direct-coupled three-stage amplifier.

8.6 OP AMPS

An *operational amplifier* (op amp) is a high-gain direct-coupled differential amplifier. (Note: The word **operational** comes from the fact that early op amps were used in analog computers to perform mathematical operations.) A *differential amplifier* has two inputs and a signal output; it depends on external feedback from output to input to determine the operating characteristics. The gain of the op amp is set with external resistors.

The schematic symbol for an op amp is shown in Figure 8.12. The op amp has two inputs, labeled inverting (−) and noninverting (+). The output is shown on the right side of the symbol in Figure 8.12. Power-supply connections are shown at the top and bottom of the symbol.

The characteristics of op amps closely approach those of an ideal amplifier. The characteristics are

◆ high-input impedance

◆ very high gain

◆ low-output impedance

◆ frequency response down to d-c

 Key Point: *Op amps are high gain amplifiers used with a feedback network to determine operating characteristics.*

IMPORTANT

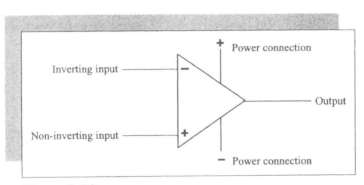

Figure 8.12
Schematic symbol for an op amp.

Self-Test

8.1 An amplifier is a device that produces an output larger than its _____ .

8.2 Name three main characteristics of amplifiers.

8.3 Unwanted changes in the output signal waveform are called _____ .

8.4 To provide maximum power transfer, the output impedance of the source should match the _____ impedance of the load.

8.5 The frequency limits of amplifier bandwidth are determined by the _____ points of the amplifier.

8.6 Variation in power supply can cause _____ in a direct-coupled amplifier.

8.7 The gain of a multistage amplifier is equal to the _____ of the gains of the individual stages.

8.8 Capacitive coupling of amplifier stages limits _____ response.

8.9 A differential amplifier amplifies the _____ between its two inputs.

8.10 Voltage gain in an op amp is the ratio of the output voltage to the _____ voltage.

Oscillators

TOPICS

Oscillator Basics
Oscillator Types

Key Terms Used in This Chapter

SINUSOIDAL	Having the shape of a sine wave.
POSITIVE FEEDBACK	A portion of an amplifier output signal that is fed back to the amplifier input to reinforce the output signal.
TANK CIRCUIT	A circuit in which an inductor and a capacitor are connected in parallel.

9.1 INTRODUCTION

In many industrial applications, a-c signals are needed to transfer information or to begin a process. These a-c signals usually are generated by electronic circuits called oscillators. Basically, an oscillator generates an a-c signal at a frequency determined by the circuit components and characteristics.

The purpose of this chapter is to introduce and discuss the basics of oscillators. Along with the basics, the chapter describes two of the most common kinds of oscillators and compares the advantages and disadvantages of each.

9.2 OSCILLATOR BASICS

Without the electronic oscillator, very few advanced electronic circuit applications would be possible. An *oscillator* is a circuit that produces a continuous output signal. Thus, the purpose of an oscillator is to perform as a **signal generator**. More specifically, the primary function of an oscillator is to generate a given waveform at a constant amplitude and specific frequency and to maintain this waveform within certain limits. The oscillators described in this chapter will produce a sinusoidal waveform. While it is true that a mechanical alternator will also produce a sinusoidal waveform at a given frequency (such as 60 Hz from an a-c power line), and it is possible to maintain the frequency within very strict limits, it is also true that alternators are *not* oscillators. Before the advent of the transistor, some alternatives were made that could produce a radio frequency, but even their top frequency limit was still very low in comparison to RF. Along with the disadvantage of their frequency limitations, they were very expensive and could be used at only one frequency. The development of the transistor (and vacuum tube), however, solved the problem of generating high-frequency a-c.

So what is an oscillator?

By definition, an oscillator is a non-rotating device (this eliminates the alternator, which is a rotating device) for producing alternating current, the output frequency of which is determined by the characteristics of the device.

We can enlarge upon the basic definition in the following manner: *An oscillator is a passive electronic component, or group of passive electronic components, that when supplied with a source of energy in the proper phase and of sufficient amplitude to overcome circuit losses, will furnish an electrical periodic function repetitively.* Because transistors are amplifiers, they may be used in oscillator circuits to provide the energy required.

Can any circuit that is capable of oscillating be a useful oscillator?

In order to qualify as an oscillator, a circuit must be able to generate sustained oscillations in a desired and controllable manner. The undesired oscillation in a public address system (loud whistles) that occurs when the microphone is moved too close to the speaker is a good example.

A transistor oscillator is essentially an amplifier that is designed to supply its own input through a *feedback* system. This feedback network must meet two conditions if sustained oscillations are to be produced. First, the voltage feedback from the output circuit must be in phase with the excitation voltage at the input circuits of the amplifier; in other words, the feedback must aid the actions taking place at the input. This is called positive (or regenerative) feedback. Second, the amount of energy feedback must be sufficient to compensate for circuit losses.

Feedback may be accomplished by inductive, capacitive, or resistive coupling between the output and input circuits. Various circuits have been developed to produce feedback with the proper phase and amplitude. Each of these circuits has certain characteristics that make its use advantageous under given circumstances.

 Key Point: *To oscillate is to move back and forth. The periodic back-and-forth motion is called oscillation. Many kinds of oscillator are sinusoidal. Sinusoidal motion has the shape of a sine wave.*
IMPORTANT

For an oscillator used in electronic devices, what are the desired characteristics? Virtually every piece of equipment that utilizes an oscillator necessitates two main requirements for the oscillator: (1) amplitude stability and (2) frequency stability.

Amplitude stability refers to the ability of the oscillator to maintain a constant amplitude output waveform—the less the deviation from a predetermined amplitude, the better the amplitude stability.

Frequency stability refers to the ability of the oscillator to maintain the desired operating frequency—the less the oscillator deviation from the operating frequency, the better the frequency stability.

The rigidity with which these requirements must be met depends on the accuracy demanded of the equipment.

9.2.1 TANK CIRCUIT

Figure 9.1 shows a parallel resonant (tuned) LC circuit called a *tank circuit*. Recall that the voltage across a capacitor cannot change instantly, nor can the current through an inductor change instantly. If the capacitor is fully charged, all the energy in the tank circuit is stored in the capacitor. The capacitor then discharges into the inductor, transferring the energy

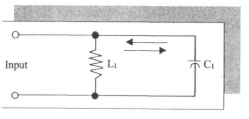

Figure 9.1
Tank circuit.

stored in the capacitor to the magnetic field of the inductor until all the energy is in the inductor.

The process of energy transfer then reverses. The energy transfers from the inductor back to the capacitor. This back-and-forth transfer of energy is called the *flywheel effect* and results in sine wave current and voltage changes in the tank circuit.

If the capacitor and inductor were ideal components, energy transfer and oscillation would continue indefinitely. However, in any electrical circuit, there exists some inherent resistance. This resistance will constitute a power loss that will eventually stop the oscillations in the tank circuit. The output of a tank circuit in which the amplitude of oscillations decreases to zero is called a **damped waveform**. Damping can be overcome only by re-energizing the tuned circuit at a rate comparable to that at which it is being used. This re-energizing process is usually accomplished through the use of a transistor acting as a switching device.

Key Point: *The oscillator circuit is essentially a closed loop utilizing d-c power to maintain a-c oscillations.*

IMPORTANT

9.2.2 FEEDBACK

Earlier, we pointed out that to convert an amplifier to an oscillator, a portion of the output signal must be fed back to the input. The feedback signal must have the correct phase to induce oscillations. Generally, the

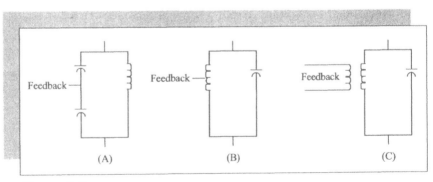

(A)   (B)   (C)

Figure 9.2
Three main feedback methods.

feedback is taken from the tuned load. Three methods are employed. These are shown in Figure 9.2. In all three types, between 10 and 50% of the output must be used as feedback.

9.3 OSCILLATOR TYPES

There are several different types of oscillators. However, in this basic text, for ease of explanation and to enhance understanding, we only discuss the Colpitts and Hartley oscillators.

9.3.1 COLPITTS OSCILLATOR

The *Colpitts oscillator*, shown in Figure 9.3, is used to produce a sine-wave output of constant amplitude and fairly constant frequency within the RF range and occasionally within the audio range. The circuit is generally used as a local oscillator in receivers, as a signal source in signal generators, and as a variable frequency oscillator for general use over the low, medium, and high frequency ranges.

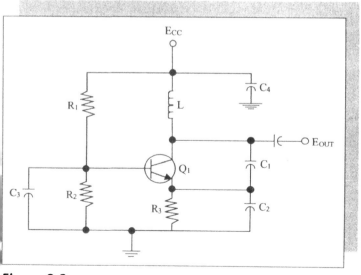

Figure 9.3
Colpitts oscillator.

The Colpitts oscillator is the most common kind of LC oscillator. The LC circuit is used to establish its frequency of operation. Inductive rather than capacitive tuning is employed, and feedback is obtained through a capacitive voltage divider circuit. Its frequency stability is good. The Colpitts oscillator oscillates easily at the higher frequencies.

9.3.2 HARTLEY OSCILLATOR

In comparing the Colpitts oscillator shown in Figure 9.3 with the *Hartley oscillator* shown in Figure 9.4, it should be obvious that they are very similar in that both of them make use of LC tank circuits. Notice in Figure 9.4 that the Hartley circuit uses the tapped primary of T_1 as the inductance. The output of the oscillator is taken from the transformer's secondary winding. This points out the major difference between the Colpitts and the Hartley; that is, the Colpitts oscillator uses a capacitive voltage divider, and the Harley uses a tapped inductor to provide the feedback.

In the past, the Hartley oscillator was used in low audio-frequency oscillators. Today, Hartley oscillators are used less because of other developments in technology, including low-cost crystals for fixed-frequency applications and electronically tunable oscillators.

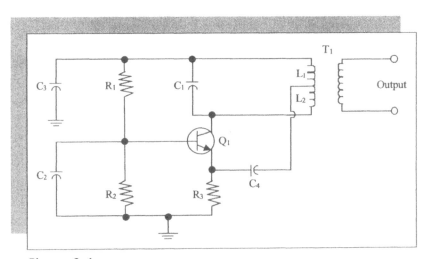

Figure 9.4
Hartley oscillator.

Self-Test

9.1 An oscillator is an amplifier with a _____ feedback network.

9.2 An _____ oscillator is a resonant circuit.

9.3 A tank circuit uses an inductor and capacitor connected in _____.

9.4 In each cycle of an LC tuned circuit, some _____ is lost due to the resistance of the inductor and capacitor.

9.5 To provide feedback, a Colpitts oscillator uses a _____ voltage divider, but a Hartley oscillator uses a tapped _____.

9.6 Two oscillators that use LC tank circuits are _____ and _____.

9.7 For parallel resonance, currents I_L and I_C are _____ and equal.

9.8 When a tank circuit oscillates, some energy is lost to _____.

9.9 What type of oscillator uses a tap on the coil for the feedback voltage?

9.10 What makes an amplifier into an oscillator?

Power Supplies

10

TOPICS

The Power Supply: What Is It?
Power Smoothers

Key Terms Used in This Chapter

POWER SUPPLY	Used to convert a power source (such as the a-c mains) to a useful form (a circuit that provides a direct-current output at some desired voltage from an a-c input voltage).
UPS	Uninterruptible power supply.
RECTIFIER	A device that allows current flow in only one direction.
INVERTER	A circuit that changes direct current to alternating current.
RIPPLE	Small voltage fluctuations in the output of a filter.
POWER SMOOTHER	A circuit or device that improves the quality of input power.

10.1 INTRODUCTION

In electronics, power supplies perform two important functions: (1) they provide electrical power when no other source is available; and (2) they convert available power into power that can be used by electronic circuits. For example, power supplies can be used to provide the d-c supply voltage needed for an amplifier, oscillator, or other electronic device. Note

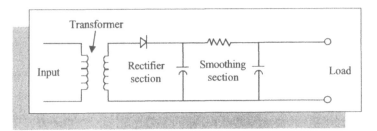

Figure 10.1
Basic power supply.

that conversion from one type of power supply (a-c to d-c, for example) does not improve the quality of the input power, so power conditioners are added for smoothing. In this chapter, we briefly discuss different kinds of power supplies and conditioners (smoothers) and how they are used.

10.2 THE POWER SUPPLY: WHAT IS IT?

A simple diagram (see Figure 10.1) provides the best way to demonstrate the actual makeup of a power supply. A basic power supply consists of four sections: a transformer (see Chapter 11), a rectifier, a smoothing section, and a load. The transformer converts the 120-V a-c to a lower a-c voltage. The choice of this a-c depends on the final level of d-c required. The rectifier section is used to convert the a-x input to d-c. Unfortunately, the d-c produced is not smooth d-c but instead is pulsating d-c. The smoothing, or conditioning section, functions to take the pulsating d-c and convert it to a pure d-c with as little a-c ripple as possible. The smoothed d-c is then applied to the load.

10.2.1 FUNCTIONS OF POWER SUPPLIES

Electrical components must have electrical power to operate, obviously. This electrical power is usually direct current. Components typically use a single low voltage, usually 5 volts, and a single polarity, either positive (+5 V) or negative (−5 V). Circuits often require several voltages and both polarities, typically +5 V, +12 V, and −12 V.

Power supplies are circuits that supply the necessary electrical power for the variety of conditions at which electronic equipment operates. For example sometimes no power is available—this may be a normal condition. Or, maybe the operating power is turned off. There could also be a main power failure. On the other hand, power might be readily available, but one or more of its characteristics—voltage, frequency, or polarity—is wrong for the application.

How about when no power is available under normal conditions—what kind of power supply is available? In this situation, the power supply is usually some kind of stored electrical energy (such as a battery) or generated electrical energy (such as that provided by a solar cell). Moreover, stored energy from capacitors can provide power for a short time.

Have you ever operated a computer when all of a sudden electrical power is lost? Along with the power loss, a considerable amount of data might also be lost. Have you been there? Well, if you have, you probably have come to appreciate the use of or the need for a backup power supply. Typically, this backup supply is provided by a *UPS (uninterruptible power supply)*. A typical UPS system consists of a bank of continuously charging batteries with enough capacity to maintain operation while an orderly shutdown, a transfer to backup power, or a restoration of main power can be effected.

Even when power is normally available and is seldom interrupted, there still may be a problem; that is, the power may not have the right characteristics for electronic devices. Consider main power, it is almost universally a-c at 60 (in the U.S.) to 50 Hz (elsewhere). The problem is that most electronic devices usually require d-c or a-c at much higher frequencies. Then, there is an additional problem with typical main power supplies. They usually range from about 100 to 240 V (120 V in the U.S.), which is much too high for direct operation of electronic circuits. *Power supplies* provide required conversions including (1) a-c to d-c; (2) a-c to a-c; (3) d-c to d-c; and (4) d-c to a-c.

Key Point: *Power supplies perform the necessary conversions.*
IMPORTANT

To this point, we have established the fact that power supplies perform the necessary conversions, but it is also important to recall what we stated earlier: power supplies, by themselves, do not improve the quality of the main power. Consider, for example, the standard household electrical power supply (in the U.S.)—it may vary from 100 to 130+ V a-c, and appliances used in the household may be labeled for 110, 112, 115, 117, or 120 V a-c. Keep in mind that the 60 Hz frequency also varies slightly. These variations are not suitable for optimum operation of electronic devices.

The problem is that if we take a simple power supply and connect it to household a-c supply, undesirable input variations at a different output level are repeated (copied). In addition, they also produce an unwanted **ripple** in their d-c output.

What is ripple? To actually view ripple, one would need an oscilloscope. On the oscilloscope, ripple appears as a wave instead of a straight line. Is the ripple wave alternating current? No; it always has the same polarity. On some electronic devices (such as radios), the effect of ripple may be negligible, but computers are seriously affected and may become inoperable if the quality of main power is not improved. Accordingly, power supplies for electronic equipment (such as computers) have smoothing networks (power conditioners) either built-in or added to them to provide stable and regulated output from the varying input.

Power supplies are composed of individual (discrete) components, such as transistors, fuses, diodes, rectifiers, transistors, resistors, and capacitors. These individual components may be wired together, the functions of an integrated circuit (IC), or both.

Power-smoothing networks are combinations of circuits that provide voltage regulation, ripple filtering, current regulation or limiting, and frequency regulation. Power-smoothing networks are incorporated into, or added to, power supplies for sensitive, critical applications.

 Note: *The term power supply is used for both simple convert-only devices and sophisticated power-smoothing devices.*
IMPORTANT

10.2.1.1 POWER SUPPLY: D-C TO D-C

One of the simplest power supplies, a d-c-to-d-c power supply, is shown in Figure 10.2. In this simple circuit, if the current drawn by the load does not change, the voltage across the resistor (often referred to as a *dropping resistor*) will be as steady as the source voltage.

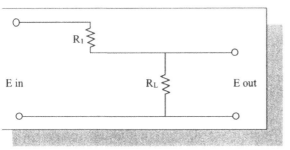

Figure 10.2
Basic d-c-to-d-c power supply.

10.2.1.2 POWER SUPPLY: A-C TO A-C

The most common way to convert a-c voltage levels is with a trans-
ormer. A *transformer* is a device that allows energy to be transferred in
n a-c system from one circuit, called the **primary**, to a second circuit,
alled the **secondary**. The function of the transformer is to deliver an out-
ut voltage that is higher or lower than the input voltage.

Compared to the dropping resistor technique discussed in Section
0.2.1.1, a transformer has several advantages and disadvantages. On the
ositive side, a transformer is very efficient due to a very low power loss,
ven with a large voltage difference between the primary and the sec-
ndary. No electrical connection is needed between the primary and the
econdary, because power is transferred by the magnetic field, which elec-
rically isolates the transformer from its output.

On the negative side, transformers are relatively large (bulky) and
eavy. They produce heat and vibration. (Note: Transformers are covered
n greater detail in Chapter 11.)

10.2.1.3 POWER SUPPLY: A-C TO D-C

The most common power supply is a-c to d-c. This arrangement uses
rectifier, which changes the 60-Hz a-c input voltage to fluctuating, or
ulsating, d-c output voltage (see Figure 10.3). The diode allows current
n only one direction, for one polarity of applied voltage. Thus, current
lows in the output circuit only during the half-cycles of the a-c input
oltage that turn the diode on.

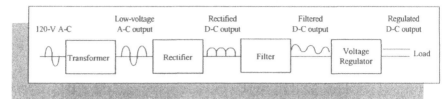

Figure 10.3
An a-c-to-d-c power supply.

The obvious question is, would the pulsating d-c output from a recti
fier be suitable for use by the load connected to it (e.g., an oscillator or
amplifier)? No, not really. Figure 10.3 shows a diagram of a typical a-c
to-d-c power supply. Notice that the input is 120-V a-c, which is typical
of the voltage supplied to most households and businesses from a utility
line. The 120-V a-c is fed to the primary of a transformer. The transformer
reduces the voltage of the a-c output from the secondary. The transformer
output is the input to the rectifier, which delivers a pulsating d-c output
The rectified output is the input to the filter, which smoothes out the pulses
from the rectifier. The filter output is the input to the voltage regulator
which maintains a constant voltage output, even if the power drawn by the
load changes. The voltage regulator output is then fed to the load.

10.2.1.4 POWER SUPPLY: D-C TO A-C

Circuits that change d-c to a-c are called *inverters*. Normally, inverters are used if an a-c voltage is needed and only d-c is available (i.e., used
in battery-powered equipment such as automobiles, boats, and airplanes)
In addition, inverters are also used to operate handheld electrical devices
to computers.

10.3 POWER SMOOTHERS

Power smoothers, or conditioners, are built into or added to power
supplies to regulate and stabilize the power supply. They include filter
and voltage regulators.

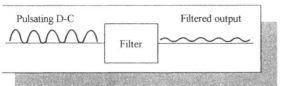

Figure 10.4
Ripple in filter output.

10.3.1 FILTERS

Filters are used in both d-c output and a-c output power supplies. In d-c output power supplies, filters help smooth out the rectifier output pulses. The key word here is help—help to filter out output pulses; they are not capable of smoothing out the pulses completely—the output has a distinct ripple, as shown in Figure 10.4. In a-c output power supplies, filters are used to shape waveforms (i.e., they remove the undesired parts of the waveform).

Key Point: *The primary function of an electronic filter is to remove an unwanted frequency or band of frequencies.*
IMPORTANT

10.3.2 VOLTAGE REGULATORS

Analogous to an automatic pressure-regulating valve on a water line, *voltage regulators* (see Figure 10.5) are designed to maintain a certain voltage automatically. Voltage regulators usually are located between the power supply filter and the load.

Key Point: *Electronic voltage regulators function to keep output voltage of an a-c-to-d-c power supply constant regardless of changes in input voltage or output current.*
IMPORTANT

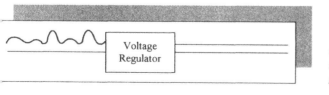

Figure 10.5
Effect of voltage regulator.

Self-Test

10.1   The use of a _____ in series with a load is a common means of providing a simple d-c-to-a-c power supply.

10.2   The part of an a-c-to-d-c power supply that converts a-c voltage to pulsating d-c voltage is the _____.

10.3   In a typical a-c-to-d-c power supply, the _____ from the rectifier is the input to the filter.

10.4   The power smoother that adjusts output voltage to compensate for load changes is the _____.

10.5   Voltage regulators usually are located between the power supply _____ and the _____.

Transformers

11

TOPICS

Transformer Construction
Transformer Operation
Transformer Nomenclature
Current in a Transformer
Autotransformers
Transformer Efficiency
Impedance Matching

Key Terms Used in This Chapter

AUTOTRANSFORMER	A single, tapped, winding used to step up or step down voltage.
EFFICIENCY	Ratio of power output to power input.
MUTUAL INDUCTANCE	Ability of one coil to induce voltage in another coil.
TRANSFORMER	A device that has two or more coil windings used to step up or step down a-c voltage.
TURNS RATIO	A comparison of turns in primary and secondary for a transformer.

11.1 INTRODUCTION

As the name implies, *transformers* are used to transform an a-c voltage to a higher or lower level.

Note: *It would be just as accurate to define a transformer as a device that allows energy to be transferred in an a-c system from one* IMPORTANT *circuit (the primary circuit) to another (the secondary circuit).*

Why is this "transformation" desired or, in many cases, necessary? Simply put, it is cheaper to transmit electric energy at high voltages rather than at low voltages because the loss of power in the line at high voltages is smaller. For this reason, the 220 V that is delivered by an a-c generator may be increased, or stepped up, to 2200 or even 220,000 V for transmission over long distances. At its destination, the voltage is stepped down to 240 V for industrial users and to 120 V for ordinary home and power users. The changes in the voltage continue. For example, in a television set, one transformer produces 28-V from the 120-V power line; and another produces several thousand volts from a 28 V oscillator. Elsewhere, the 120-V supply is reduced to 20 V or less to operate a toy train or to 12 V to operate a doorbell.

Transformers are found in communications equipment and digital equipment, where often they are the size of a postage stamp or smaller.

The communication uses of transformers are usually to isolate items of equipment to eliminate electrical noise and other static interference.

In this chapter, we cover transformer construction, operation, nomenclature, and current and voltage relationships.

11.2 TRANSFORMER CONSTRUCTION

As shown in Figure 11.1, a transformer consists of (1) the **primary coil**, which receives energy from an a-c source, (2) the **secondary coil**, which delivers energy to an a-c load, and (3) a **core** on which the two

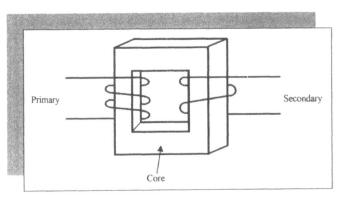

Figure 11.1
Basic transformer.

coils are wound. The core is generally of some highly magnetic material, although ceramics, cardboard, and other nonmagnetic materials are used for the cores of some electronic devices.

11.3 TRANSFORMER OPERATION

(Note: As explained in the following, the transformer is an important application of **mutual inductance.**)

An alternating current will flow when an a-c voltage is applied to the primary coil of a transformer. This current produces a field of force that changes as the current changes. The changing magnetic field is carried by the magnetic core to the secondary coil, where it cuts across the turns of that coil. These cuttings induce a voltage in the secondary coil. In this way, an a-c voltage in one coil is transferred to another coil, even though there is no electrical connection between them (mutual inductance). The number of lines of force available in the primary is determined by the primary voltage and the number of wire turns on the primary—each turn producing a given number of lines. Now, if there are **many turns on the secondary**, each line of force will cut **many turns** of wire and **induce a high voltage**. If the **secondary contains only a few turns**, there will be few cuttings and a **low induced voltage**.

The turns relationship is known as the *turns ratio*. That is, the ratio of the number of turns in the primary to the number in the secondary is the turns ratio of the transformer:

$$\textbf{Turns ratio} = \frac{\textbf{N}_P}{\textbf{N}_S} \qquad (11.1)$$

For example, 400 turns in the primary and 20 turns in the secondary provide a turns ratio of 400/20, or 20:1.

The secondary voltage, then, depends on the number of secondary turns as compared with the number of primary turns. If the secondary has twice as many turns as the primary, the secondary voltage will be twice as large as the primary voltage. If the secondary has three times as many turns as the primary, the secondary will be three times as large as the primary voltage. If the secondary has half as many turns as the primary, the secondary voltage will be one-half as large as the primary voltage, and so on. Thus, the voltage ratio is in the same proportion as the turns ratio:

$$\frac{E_P}{E_S} = \frac{N_P}{N_S} \qquad (11.2)$$

where

E_P = voltage on primary coil
E_S = voltage on secondary coil
N_P = number of turns on primary coil
N_S = number of turns on secondary coil

IMPORTANT
Important Note: Equations (11.1) and (11.2) apply only to iron-core transformers with unity coupling. Air-core transformers for RF circuits are generally tuned to resonance. In this case, the resonance factor is considered instead of the turns ratio.

11.4 TRANSFORMER NOMENCLATURE

A voltage ratio of 1:4 (read as 1 to 4) means that for each volt on the primary, there are 4 V on the secondary. This is called a *step-up* transformer. A step-up transformer **receives a low voltage** on the primary and **delivers a high voltage** from the secondary. A voltage ratio of 4:1 (read as 4 to 1) means that for 4 V on the primary, there is only 1 V on the secondary. This is called a **step-down** transformer. A step-down transformer **receives a high voltage** on the primary and **delivers a low voltage** from the secondary.

11.5 CURRENT IN A TRANSFORMER

In the efficient transformer, the power delivered to the primary is transferred to the secondary with practically no loss. For all practical purposes, the power input to the primary is equal to the power output of the secondary, and the transformer is assumed to operate at an efficiency of 100 percent. Thus,

$$\text{Power input} = \text{Power output} \qquad (11.3)$$

Because

$$\text{Power input} = E_P \times I_P \qquad (11.4)$$

and

$$\text{Power output} = E_S \times I_S \qquad (11.5)$$

$$E_P \times I_P = E_S \times I_S \qquad (11.6)$$

By dividing both sides of the equation by $E_S \times I_P$ and canceling identical terms, we obtain the equation

$$\frac{E_P}{E_S} = \frac{N_P}{I_P} \qquad (11.7)$$

This equation indicates that the current ratio in a transformer is **inversely proportional** to the voltage ratio; that is, voltage step-up in the secondary means current step-down, and vice versa. Stated another way, if the **voltage** ratio **increases**, the **current** ratio will **decrease**. If the **voltage** ratio **decreases**, the **current** ratio will **increase**. The secondary does not generate power but only takes it from the primary.

 Important Point: *When making transformer calculations, remember that the side with the higher voltage has the lower current.* IMPORTANT *The primary and secondary voltage and current are in the same proportion as the number of turns in the primary and secondary.*

11.6 AUTOTRANSFORMERS

As illustrated in Figure 11.2, an *autotransformer* consists of one continuous coil, one single winding. This winding may be tapped at any point along its length to provide a set of three terminals (1, 2, 3). The winding 1-2 is primary, and the entire winding 1-3 is the secondary. Even though

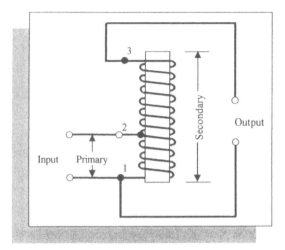

Figure 11.2
A simple autotransformer.

the windings have an electrical connection at 2, the principle of operation remains the same. When an a-c voltage is applied to the primary winding 1-2, the lines of force around 1-2 link the turns between 1 and 3, inducting a higher or lower voltage according to the formula ER = TR.

Autotransformers are used often because they are compact, efficient, and usually cost less with only one winding. However, the same size wire must be suitable for both the primary and secondary.

11.7 TRANSFORMER EFFICIENCY

Efficiency is defined as the ratio of power out to power in. Stated as an equation,

$$\textbf{Efficiency} = \frac{P_{out}}{P_{in}} \qquad (11.8)$$

Thus, a transformer that delivers *all* the power put into it would have an efficiency of 100%. Actually, because of copper and core losses, the efficiency of even the best transformer is less than 100%. In a transformer that is less than 100% efficient, the primary supplies more than the secondary power. The primary power missing from the output is dissipated as heat in the transformer.

11.8 IMPEDANCE MATCHING

The maximum transfer of energy from one circuit to another will occur when the impedances of the two circuits are equal or **matched**. If the two circuits have unequal impedances, a coupling transformer may be used as an intermediate impedance-changing device between the two circuits. By constructing the transformer's winding so that it has a definite turns ratio, the transformer can perform any impedance-matching function. The turns ratio establishes the proper relationship between the ratio of the primary and secondary winding impedances.

Self-Test

11.1 How is a transformer constructed?

11.2 What is used as an input to a transformer?

11.3 What is meant by the turns ratio?

11.4 If E_{in} = 10 V_{pp} and E out = $7E_{pp}$, what is the turns ratio?

11.5 What is the efficiency of a transformer if it draws 900 watts and delivers 600 watts?

Final Review Examination

The answers for the comprehensive examination are contained in Appendix B.

12.1 What are the four basic parts of an electrical circuit?

12.2 The physical size of a resistor has no relationship to its _____.

12.3 If the current through a conductor is doubled and the resistance is constant, the power consumed by the conductor will increase to _____ times the original amount.

12.4 What is the total resistance of three 30-Ω resistors connected in series?

12.5 Eight 10-Ω resistances are in series across a 120-volt source. What is the voltage drop across each resistance?

12.6 The equivalent resistance R_T of parallel branches is _____ than the smallest branch resistance because all the branches must take _____ current from the source than any one branch.

12.7 When two resistances are connected in parallel, the voltage across each is the _____.

12.8 If each of two resistances connected in parallel dissipates 10 watts, the total power supplied by the voltage source equals _____ watts.

12.9 If an a-c voltage wave has an instantaneous value of 90 volts at 30°, find the peak value.

12.10 A transmitter tuning coil must have a reactance of 95.6 Ω at 3.9 MHz. Find the inductance of the coil.

12.11 What is the total capacitance of three capacitors connected in parallel if their values are 0.15 μF?

12.12 What charge (Q) is taken on by a 0.5-F capacitor connected across a 50-volt source?

12.13 Find the resonant frequency of a series circuit if the inductance is 300 μh and the capacitance is 0.005 μF.

12.14 A transformer whose efficiency is 80% draws its power from a 120-volt line. If it delivers 192 watts, find the power input.

12.15 A transformer transmits energy from one circuit to another by means of electromagnetic _____ between the two coils.

12.16 Reactance decreases as the frequency or capacitance _____.

12.17 A component called a _____ is most commonly used as a rectifier.

12.18 Actual electron flow is from _____ to _____, the opposite of conventional current.

12.19 A diode readily conducts electricity if it is _____-biased.

12.20 Holes move in the same direction as _____ current.

12.21 A _____-biased diode conducts electricity.

12.22 The _____ region in a junction diode has very few electron-hole pairs.

12.23 The _____ transistor amplifier connection provides the highest power gain.

12.24 Another name for an op amp is _____.

12.25 Amplifier biasing components are typically _____.

12.26 An emitter-follower amplifier is used when a high input _____ is desired.

12.27 Power amplifiers have high _____ and high _____ output.

12.28 Amplifier _____ expresses output power as a percentage of d-c input power.

12.29 The frequency limits of amplifier bandwidth are determined by the _____ points of the amplifier.

12.30 Variations in power supply can cause _____ in a direct-coupled amplifier circuit.

12.31 An _____ is an amplifier with a positive feedback network.

12.32 An LC oscillator circuit is a _____ circuit.

12.33 State the law of attraction and repulsion of charged bodies.

12.34 In a series LCR circuit, under what specific conditions is impedance minimum and line current maximum?

12.35 Why is positive feedback necessary rather than negative feedback in an oscillator?

12.36 What feedback method is used in a Hartley oscillator?

12.37 If $R = 10$ kΩ and $C = 1$ μF, find the time constant.

12.38 What feedback method is used in a Colpitts oscillator?

12.39 In the FET, what controls the flow of current?

12.40 What is the relationship between X_L and X_C at resonant frequency?

Appendix A

Answers to Chapter Self-Tests

CHAPTER 2

2.1   24

2.2   18

2.3   20

2.4 $x = 1$

2.5 $x = 0.5$

2.6 $x = 2$

2.7 $y = 3$

2.8 $y = 2$

2.9 $R = P/I^2$

2.10 $E = IR$

CHAPTER 3

3.1   (a) Volt, E; (b) amp, I; (c) coulomb, Q; (d) ohm, R

3.2 E and R

3.3   Value of resistors, length of conductors, and the diameter of the conductors

3.4   Make wires longer or decrease their diameter.

3.5 $P = E \times I$

200 w = 110 $V \times$?

200/110 = 1.8 amp

3.6 Series has one path for current flow, while a parallel circuit has more than one path.

3.7 Source voltage

3.8 $C_T = C_1 + C_2 + C_3 + \cdots + C_n$

$C_T = 120\ \mu F + 120\ \mu F + 120\ \mu F = 360\ \mu F$

3.9 The time required to charge a capacitor to 63% of its maximum value or to discharge it to 37% of its maximum value

3.10 TC = RC

= 2 MΩ × 2 μF

= 4 s

CHAPTER 4

4.1 A voltage applied across a semiconductor junction so that it will tend to prevent current flow

4.2 A voltage applied across a semiconductor junction so that it will tend to produce current flow

4.3 Positive, negative; negative, positive

4.4 Holes

4.5 Stop

4.6 Very high

4.7 N-type and P-type

4.8 Free electrons in the N section move away from the junction.

4.9 Depletion region

4.10 To decrease its resistivity

CHAPTER 5

5.1 The ratio of collector current to base current

5.2 Voltage on the gate.

5.3 High

5.4 Forward

5.5 Voltage

5.6 Gate

5.7 Depletion

5.8 High, high

5.9 On

5.10 Current

CHAPTER 6

6.1 The conductor(s) must cut a maximum number of lines of force.

6.2 A voltage is induced in the conductor.

6.3 A waveform that represents all the instantaneous values of current or voltage as the armature of an alternator moves through one complete rotation.

6.4 Magnitude and direction

6.5 846 volts (600 × 1.41)

6.6 a-c, cut, counter

6.7 Current has an associated magnetic field.

6.8 Number of turns in the coil; length of the coil; cross-sectional area of the coil; permeability of the core

6.9 Decrease

6.10 Area of plates, distance between plates, dielectric constant

6.11 The time required to charge a capacitor to 63% of its maximum value or to discharge it to 37% of its maximum value

6.12 Charge

6.13 a-c

6.14 Inductance; frequency; 2π

6.15 Voltage lags current by 90°.

6.16 Average power actually consumed by an a-c circuit.

6.17 Average power

6.18 90

6.19 1

6.20 Ratio of average power to apparent power

CHAPTER 7

7.1 Maximum output

7.2 Minimum output

7.3 $Z = $ minimum

7.4 Resonant frequency

7.5 Maximum, equal

7.6 Maximum

7.7 Minimum

7.8 0

7.9 The width of the resonant band of frequencies centered around f_r

7.10 Resistance

CHAPTER 8

8.1 Input

8.2 Gain, bandwidth, distortion

8.3 Distortion

8.4 Input

8.5 Half-power

11.3 The ratio of turns in the primary winding to the number of turns in the secondary

11.4 $TR = 1.43{:}1$

11.5 $\mathrm{Eff} = P_s/\mathrm{pp}$
$$= 600/900$$
$$= 0.667 = 66.7\%$$

Appendix B

Answers to Chapter 12—Final Review Examination

12.1 Voltage source, conductor, load, control devices

12.2 Resistance

12.3 4 (I^2R)

12.4 90 Ω

12.5 15 volts

12.6 Less, more

12.7 Same

12.8 20 watts

12.9 90 V $= E_m \sin(30°) = E_m (0.5) = 180$ volts

12.10 3.9 mh

12.11 0.45 μF

12.12 $Q = 25$ C

12.13 130 Khz

12.14 240 watts

12.15 Induction

12.16 Increases

12.17 Diode

12.18 Negative, positive

12.19 Forward

12.20 Conventional

12.21 Forward

12.22 Depletion

12.23 Common emitter

12.24 Differential amplifier

12.25 Resistors

12.26 Impedance

12.27 Voltage, current

12.28 Efficiency

12.29 Half-power

12.30 Drift

12.31 Oscillator

12.32 Resonant

12.33 Like charges repel each other and unlike charges attract each other.

12.34 When the circuit is at resonance

12.35 Positive feedback will cause the amplifier to sustain an oscillator. Negative feedback will cause amplifier to stabilize, which decreases oscillations.

12.36 An inductive voltage divider

12.37 $T = 0.01$ seconds

12.38 A capacitive voltage divider

12.39 The voltage in the gate controls the flow of drain current.

12.40 $X_L = X_C$

Index